WORLD STATE OF EMERGENCY

This book is dedicated to my dear friend, Shahin Nezhad,
the *Shahriyâr* of the Iranian Renaissance,
and to his wife, Artemis, who is worthy of her name.

JASON REZA JORJANI

World State of Emergency

ARKTOS
LONDON 2017

Copyright © 2017 by Arktos Media Ltd.

All rights reserved. No part of this book may be reproduced or utilised in any form or by any means (whether electronic or mechanical), including photocopying, recording or by any information storage and retrieval system, without permission in writing from the publisher.

ISBN	978-1-912079-93-3 (Softcover)
	978-1-912079-91-9 (Ebook)
COVER AND LAYOUT	Tor Westman

www.arktos.com

CONTENTS

Introduction — vii

1. The Third World War — 1
2. Planetary Emergency — 35
3. The Neo-Eugenic World State — 69
4. Robotics and Virtual Reality — 107
5. Persian Gulf of the 21st Century — 125
6. Aryan Imperium (*Iran-Shahr*) — 153
7. The Indo-European World Order — 193

Bibliography — 207

Index — 213

" … Inexorably, hesitantly, terrible as fate, the great task and question is approaching: how shall the earth as a whole be governed? And to what end shall 'man' as a whole — and no longer as a people, a race — be raised and trained?"

— Friedrich Nietzsche, *The Will to Power*

Introduction

Apocalypse. In its original Greek sense, the word means "revelation." Over the course of the next several decades, within a single generation, certain convergent advancements in technology will reveal something profound about human existence. Biotechnology, robotics, virtual reality, and the need to mine our Moon for energy past peak oil production, will converge in mutually reinforcing ways that shatter the fundamental framework of our societies. It is not a question of incremental change. The technological apocalypse that we are entering is a Singularity that will bring about a qualitative transformation in our way of being. Modern Western sociopolitical systems such as universal human rights and liberal democracy are woefully inadequate for dealing with the challenges posed by these developments.

The technological apocalypse represents a *world state of emergency*, which is my concept for a *state of emergency* of global scope that also demands the establishment of a *world state*. An analysis of the internal incoherence of both universal human rights and liberal democracy, especially in light of the societal and geopolitical implications of these technologies, reveals that they are not proper political concepts for grounding this world state. Rather, the planetary emergency calls for worldwide sociopolitical unification on the basis of a deeply rooted tradition with maximal evolutionary potential. This living heritage that is to form the ethos or constitutional order of the post-national world state, on an existential level, is the Aryan or Indo-European tradition shared by the majority of Earth's great nations — from Europe and the Americas, to Eurasia, Greater Iran or the Persianate world, Hindu India, and the Buddhist East.

The term Aryan (or *Airyânâ*, *Irân*, or *Erin*) literally means "well put together" or "finely wrought", in other words "crafty", and only derivatively "noble" for this reason. This may have been a term that others first applied to the Indo-Europeans or that they adopted in the face of their encounter with cultures that utterly lacked their crafts mastery and industrious genius. The term Aryan is also associated with the skillful path to Wisdom or *Prajnâ* in the Buddha Dharma. The native name of Buddhism, given to it by Siddhartha Gautama himself, is *Âryâ Ashtangâ Marga* or the "Aryan Eightfold Path", which is fundamentally based on *Âryâ Chatvâri Satyâni* or the "Aryan Fourfold Truths" (often loosely translated into English as the "Four Noble Truths"). The primary symbol of the Dharma, both in its Buddhist and Hindu varieties, is of course the *Swastikâ*.

The Indo-European World Order is one of three possible forms that the world state can take. The other two are a Chinese planetary hegemony, probably in the guise of the Shanghai Cooperation Organization, and the first truly global Islamic Caliphate, whether it evolves out of the current Islamic State and Al-Qaeda or some more moderate but far more dangerous transnational Muslim movement that displaces these groups. Although these three geopolitical blocks have hardly even recognized themselves, least of all the Indo-Europeans, they are already mutually engaged in a Third World War. This final and most decisive geopolitical battle will also be known as The Longest War. That is partly because, unlike the first and second world wars, the third is a clash of civilizations rather than a conflict between nation-states. It is the war that will move us beyond the nation-state as the basic unit of political organization. The United Nations International System formed at the end of World War II, with the *Universal Declaration of Human Rights* as its Magna Carta, is bound to be deconstructed by the kind of civilizational warfare that began on September 11, 2001.

This is the subject of my first chapter. It argues that the *Universal Declaration of Human Rights* (UDHR) is internally incoherent and self-undermining. By establishing an unqualified freedom of religion as one of the core human rights with universal applicability, the drafters disregarded the possibility that a "moral majority" in any particular country could use the fundamental precepts of their religion as a basis to deny the putative human rights of dissenting minorities. Based on a European Enlightenment political philosophy, the founders of the United Nations failed to recognize that a certain religion, such as Islam, can conflict with some of the most important human rights that these gentlemen wanted to believe are universal, while at the same time being a religion that quite consciously asserts itself to be impervious to reform or progressive evolution.

Samuel Huntington argues that in "the clash of civilizations", religion is more determinative than secular political ideologies. As the main antagonist in this civilizational clash, the worldwide Muslim *Ummah* are in possession of a religion whose holy book is also a coherent political constitution. Despite being "revealed" in 7th century Arabia, many of the articles of this constitution, including those which are most repugnant to the advocates of universal human rights, are not subject to amendment. I make that case very clearly. An Islamic World Order or Global Caliphate is one where women, worldwide, are legally subordinated to men and have very few rights, one where slavery is legalized, religious and intellectual dissent is criminalized, minorities are forced to pay tribute, and common thieves face draconian punishments like amputation.

It is important to remember that Islam does not even need to conquer the world by the force of arms. Its conquest of the Earth by demographics alone is nearly assured. By the later part of this century, Muslims will have a decisive majority on this planet. This brings us to the question of the globalization of democracy, the focus of Chapter 2. Based on current demographic trends, by 2075 a global democracy will mean an Islamic World State. Period. I argue that this shows how liberal democracy is incoherent. Liberalism is a distinct European political ideology that can easily be divorced from democracy.

In fact, democracy had extremely conservative and illiberal roots, both in classical Greece and when it was revived by Jean-Jacques Rousseau in the early modern era as a romantic antidote to rationalist Natural Rights (i.e. Human Rights) theories. Both the classical Greeks and Rousseau recognized that religion needs to be the basis of the general will that cohesively constitutes the sovereign decision-making power of a democratic polity. Rousseau even praised Islam as the best type of "civic religion" precisely because, unlike Christianity, at its very foundation it refuses to acknowledge any separation between religious authority and secular statecraft. Meanwhile, liberalism, as epitomized in its purest form by John Stuart Mill, advocates a false and untenable neutrality of the state with respect to the personal choice of citizens to live their lives in various ways. I show how Mill, in a way characteristic of liberals, sneaks in unjustified cultural prejudices about the kinds of individual conduct that are acceptable within a liberal or free society. His "principle of harm" is ill-defined and motivated by a misguided attempt to forge a political order that rejects civilizational ethos and collective identity. Liberal radical individualism is certainly radically anti-democratic.

Carl Schmitt analyzed the conflict between liberalism and democracy as part of his formulation of what he called "the concept of the political." According to Schmitt, liberals and democrats confuse the political with economic and other social structures, whereas the political as such is not open to endless debate and negotiation nor is it a matter of personal preference. The very concept of the political concerns who has the power to make a sovereign decision over the life and death of those within the state. During an emergency or state of exception, the written constitution of a state is suspended and parliament cannot hold a debate on what course of action to take. This state of emergency reveals not only who the sovereign is, but on what basis he makes decisions affecting the lives and property of his subjects. Insofar as his decision-making is effective, it reveals the true existential constitution or *ethos* of his people. This underlies the written constitution and is the basis for his folk's at least tacit acceptance of his dictates even if they violate the more superficial and derivative, legal constitution. Such a crisis may end with the restoration of the established legal order, but the state of emergency might also be a state of *emergence* wherein a new constitutional order reveals itself as a more truthful reflection of the spirit of a folk who find themselves anew.

At the core of sovereign decision-making in the state of emergency is the determination of Friend and Enemy. Carl Schmitt argues that the sovereign is the person or council that can effectively decide, in an emergency situation, who one's "own" people are, who their friends or allies are, and who the enemy is, even if that enemy is within one's own borders. Lines on a map do not define a nation. The *ethos* or existential character of a world-historical community does that. A nation can be extended across many arbitrary borders drawn on the basis of a failure to grasp political distinctions conceptually. Such a nation may have one or more enemies in its midst — enemies that threaten its constitutional order on an existential level. In accordance with this analysis, Schmitt thought that there could never be a single world state and that any attempt to appeal to a "common humanity" as a political concept was cheating, because Humanity-as-a-whole has no concrete enemy. What is worse is that universalist discourse about the political rights of Humanity or a universal democracy actually demonizes all those human beings who, on the basis of their own cultures, reject this discourse. Schmitt considered this planetary imperialism of Humanists to be the most tyrannically inhuman ideology, and one exemplified, in his time, both by the NATO alliance and the Soviet Union.

However, at the very end of his life, Schmitt began to reconsider this position. He did so with a view to dramatic technological developments that he admitted could be considered as something like a threat to all of humanity. For him, the foremost of these developments were the proliferation of miniaturized nuclear weapons in an age of de-territorialized partisan warfare (i.e. transnational terrorism), and a militarization of space that for the first time encompasses and frames the entire Earth as a theater of combat. Although he admitted that this might require rethinking the concept of the political, and its core Friend-Enemy distinction, Schmitt never really had the time or energy left to further develop these insights in his *Theory of the Partisan*. In a way, his last insights were also untimely. Only now, when we face technological developments that could allow us to modify our human form of embodiment, lose ourselves in spectacular simulacra, or face an interplanetary war with some post-human cybernetic intelligence, do we have a concrete context for thinking what still remained unthinkable to Carl Schmitt, namely a *world state of emergency*. From Chapters 3 to 5 of this book, I show how a number of technological innovations each require a global scope of sovereign decision, for the sake of the mere survival of anything recognizable as Humanity and with a view to averting a variety of horrifically inhuman futures.

Chapter 3 addresses Neo-Eugenic biotechnologies that require global regulation. Biotechnology offers us a variety of potentials to promote human flourishing. Using embryo selection, which is an augmented process of in-vitro fertilization where you are able to profile the embryos that are then reinserted in the womb for development, it is possible to eliminate hereditary diseases. It is also known that there is about a 15 point range in IQ between the brightest child one could have and the one that will have the hardest time studying various disciplines. IQ can be increased by around 15 points per generation, using embryo selection alone (without modifying the genetic structure of the embryo). Of course, "intelligence" is an imprecisely defined term. But there are certain factors of intelligence that have to do with the ability to learn Physics and manipulate complex mathematics, which have very strong genetic correlates. It is very important that we establish some kind of consensus regarding how this technology is going to be used. If we were to have a 15 IQ point increase per generation in only one country or one culture, this would introduce a dangerous imbalance of geopolitical power.

On account of the history that Eugenics has in the Western world, views about using this technology for what are essentially Neo-Eugenic purposes are largely negative in the West. By comparison, the Chinese scientific, academic, and political establishment is almost 100% in favor of using emerging biotechnologies to, for example, enhance the IQ of its population. The specific factors of intelligence that can be manipulated by this process are the very ones that lead to engineering breakthroughs, which historically, unfortunately, have had their first manifestations in military technology. Even within a single country, if only the wealthy have access to this kind of technology, we could see class disparities turn into real caste distinctions between a genetic aristocracy and others who are not so genetically fortunate.

The societal implications are even more significant when we think about genetic engineering, which actually does modify the genetic structure of an embryo. We can lengthen lifespan. It has been found that mice who have been genetically engineered for a longer lifespan, also have compressed morbidity — meaning that at the very end of their lives they decline very quickly rather than going through a prolonged aging process. We would also be able to enhance physique. The same techniques that were initially developed to treat Lou Gehrig's disease have been applied to the end of boosting muscle mass and decreasing the chances of obesity. Then, in terms of cognitive functioning, there are also genetic engineering techniques that were developed initially to treat Alzheimer's that have been used to boost memory capacity by two or three times. These particular enhancements sound very positive, but genetic engineering also gives us the capability of splicing human and animal genes. We really ought to ask ourselves whether we want to create a hybrid species. That is not a decision that ought to be left to one country or culture, let alone a corporation operating in the global free market.

One of the most controversial biotechnologies is human cloning. Even if we were to have a fairly widespread consensus that human cloning for the reproduction of identical persons (in other words, the creation of large numbers of 'twins') is not a great idea, human cloning is implicitly part of embryo selection. It is often the case that when an optimal embryo is inserted into the womb for development, the initial implantation is unsuccessful and another attempt has to be made. Consequently, if one already has arrived at an optimal embryo, cloning that embryo a number of times allows for repeated attempts at implantation. Since cloning is going to be part of augmenting embryo selection (a more acceptable form of biotechnology), there is no telling how else cloning might be used unless we have an effective regulatory system on a global scale.

The worst thing that may happen with biotechnology, which also brings us into the domain of robotics or cybernetics, is that some country or civilization might unilaterally decide to genetically engineer a race of slaves. The creators of these biomechanical robots might find a way to convince us that their creatures are not really human beings who would be the bearers of any political rights. Gene splicing that incorporates the elements of the genome of other animals might make it possible for these humanoids to work under adverse environmental conditions that would be difficult for humans to tolerate.

To an extent, biotechnology is converging with research in robotics, which I consider together with virtual reality in Chapter 4. One of the most recent advancements in robotics research is biomimetic design. This is the idea that we should look to insects and other non-human animals for inspiration in robotics. In particular, robotic spiders and robotic flies have been developed, which would be ideal for surveillance. They can crawl under your door or fly in through your window. They look like real insects and yet they are equipped with miniaturized components that allow them to provide surveillance to their remote operator. We will soon be living in a world where one cannot tell whether the insects that one is trying to swat in one's home are actually surveillance drones. More frighteningly, these robotic insects could be designed to mimic mosquitoes with stingers — except that the 'stinger' is a micro-syringe that injects poison into a targeted person. Another development that is related to this is work on transformers — robots that can shape shift, at institutions such as Carnegie Mellon. There will soon be biomimetic robots that can change from being a spider into being a fly by reorganizing their structure. So while thinking that one has successfully shoed a fly away, the same robot reenters one's home in the form of a spider that continues to offer both audio and video surveillance — or, in the worst case, carry out the most untraceable assassination. Terrorism will reach another level when drone robotic flies fitted with micro-charges can suddenly converge on a targeted person in a park and detonate their explosives near his jugular vein.

We are now witnessing a proliferation of drones, especially in the context of warfare. One of the technologies that the Pentagon is developing under DARPA is the ability to project a pilot's mind into a drone through virtual reality so that the pilot really feels like he is flying in whatever environment the drone is patrolling. Through a haptic suit, the pilot gets tactile feedback from the airframe of the drone. This begins to give the drone operator a different sense of embodiment than a human being ordinarily has. There are brain interfaces being developed that will allow the pilot to control the movements of the drone without any explicit commands. In other words, by thought alone. The aim here is to improve reaction time in aerial dogfights. But if one combines this system with fully immersive virtual reality, including tactile feedback from the airframe of the drone, are we really even dealing with a human being anymore? Will these drone pilots begin to dream of themselves as some other type of creature? The first form of posthuman Artificial Intelligence may be parasitic on the organic elements of human intelligence.

In terms of computer technology, the most challenging development is going to be virtual reality. We have had various primitive virtual reality systems or attempts at virtual reality. But within the next several decades, we are going to have fully immersive virtual worlds that we cannot tell apart from 'reality.' Today one needs head mounted displays in order to enter virtual reality. These may become very light-weight goggles or even contact lenses, and haptic suits are being developed to give full-body tactile feedback. Given how significant a problem on-line addiction is right now, and given how the age ceiling is rising amongst people who participate in immersive simulacra like *World of Warcraft*, one can imagine large segments of developed societies checking out of reality, losing touch with real people, and losing themselves in virtual worlds. Virtual reality could become the most addictive drug that there has ever been in human history.

When the first cinema films were developed, the people who sat in theaters where black and white films were projected and saw an oncoming train, actually fled toward the exit because they were afraid of being run over. Our minds slowly adjusted to the sophistication of these simulacra. But now we are going to be able to evoke virtual realities that are indistinguishable in fidelity from our ordinary experience. The fact that we have had all of these films around the turn of the millennium that call into question whether we are living in the actual world or whether we are living in a simulacrum portends a civilizational crisis. If you have a whole society whose members begin to question whether they are living in the real world, that society becomes vulnerable by comparison to others that are more grounded or well-rooted.

There are, nonetheless, all kinds of positive applications of virtual reality. It allows any surgeon to have a telepresence at a surgery for which he is uniquely qualified. In the context of long distance space travel, VR would allow people to be able to tolerate those voyages much better than they would be able to now. Virtual reality would afford miners on the Moon a greater capacity to endure their long term isolation. This is a very real consideration since, as I go on to argue in Chapter 5, the Moon is going to be our primary energy source for at least the remainder of this century if not long after that.

All of the technological innovations that will have been discussed in Chapters 3 and 4 require the persistence of a growing industrial economy. That cannot be taken for granted in light of the imminent peak in oil production and power generation based on petroleum. Chapter 5 begins by making the case that no existing alternative energy technologies can safely deliver the requisite power necessary for continued industrial development past peak oil. Then I turn to examine the socio-political implications of the one proposed alternative that certainly does fit the bill: environmentally friendly fusion power based on Helium-3 mined from the lunar surface.

In theory, it is possible to build a clean nuclear plant using fusion rather than fission to power the reactor. Instead of splitting atoms, a fusion reactor heats them to millions of degrees centigrade so that they can be fused. There is an experimental reactor of this kind running in Oxfordshire, using hydrogen as the fuel. While it produces enough power for an entire town, 80% of the energy comes out as neutron particles and damages the reactor wall if run for more than a few seconds at a time. Hydrogen could, however, be replaced with Helium-3 gas, to which it is very close in chemical composition. The reactor would only need a kilogram of this fuel to produce the same amount of energy as a power plant burning 10,000 tons of fossil fuel. The problem is that Helium-3 does not naturally occur on Earth and is only found in decommissioned nuclear weapons.

Helium-3 is a gas ejected from the sun and blown through space by solar winds. It is blocked by the Earth's atmosphere, but since there is nothing to shield the Moon from it Helium-3 is trapped by the lunar soil. Over billions of years, millions of tons of deposits of it have built up in the regolith — enough to power nuclear fusion plants on Earth for hundreds of years. Harrison Schmitt, a geological scientist on board the Apollo missions has analyzed lunar samples that contain significant quantities of Helium-3. Tranquility Base, the northern Polar region and one other area of the Moon are known to have high concentrations of it. Together with Gerald Kulcinski, who has conducted nuclear fusion experiments with Helium-3, Schmitt has founded a corporation for strip mining the moon and transporting H3 back to Earth as a liquefied gas (and no, the mining strips will *not* be visible from Earth). Kulcinski has called the Moon the "Persian Gulf of the 21st century." As he points out, if the lunar surface were lined with gold bricks it would not pay to go there to retrieve them, but at an estimated market value of several billion dollars a ton Helium-3 is well worth it. One metric ton of the material would supply a year's worth of energy for 10 million people in a large modern city.

Schmitt and Kulcinski are not the only people considering the prospect of Helium-3 lunar mining to fuel nuclear fusion plants. Nikolay Sevastyanov of RSC Energia (now S.P. Korolev Rocket and Space Corporation Energia), Russia's most successful rocket company with a proven track record of putting people into space and regularly servicing the International Space Station, is also actively planning such an enterprise. Sevastyanov believes that if you decisively set people a task, then they will solve it. He aims to begin mining the Moon for Helium-3 on an industrial scale by the 2020s. This is the same time frame for NASA's planned return to the Moon, which has in turn been at least partly prompted by China's announcement that it intends to attempt a manned mission in 2018.

These rivalries bring us to the heart of the matter. There is no agreed upon international framework adequate for governing resource acquisition on the Moon. A claim of property is a claim of political sovereignty, and the extension of geopolitical divisions from the Earth to the Moon means a militarization of space and an interplanetary scope of future warfare. As you may recall, this is one of the specters that caused Carl Schmitt to question his formulation of the concept of the political. If the

Moon, and eventually other rocky bodies throughout our solar system bearing Helium-3, are going to be home to vast mining colonies that supply our future power source back here on Earth, we really need to consider preempting the possibility of a war in space. Almost all terrestrial wars have been motivated, to some extent, by resource acquisition, but the broadening of the scope of these conflicts into the solar system re-defines the entire Earth as a distinct territory in a wider exopolitical sphere of sovereign decision.

In this sense Helium-3 fusion power poses a similar challenge as the effective global regulation of biotechnology, and in a similar time frame. Based on the pace of improvement in computer processing power, which allows for more adequate gene sequencing, we have not more than another 30 years to establish a world government. In the case of biotechnology, regardless of what people's opinions on it might be, there is no way to avoid a global caste system, or possibly even a speciation of humanity, unless we move toward a socialist economic system on a planetary scale that subsidizes approved forms of biotechnology. Equitable extraterrestrial resource distribution might also be the only way to avoid conflict on the Moon and deeper into the solar system.

However, the world state of emergency demands far more than a technocratic or procedural planetary government. If we were to have robotic insects all over the place collecting intelligence, then we essentially live in a world where there is no longer privacy. That these devices, soon widely available, could also be used for untraceable murders means that we will no longer be safe in our private spaces. Virtual reality *avatars* and *agents* also pose very serious threats to personal identity. What this means is that we need to move fast towards a maximal trust society. These kinds of technological developments discussed in Chapters 3 through 5 cannot take place without resulting a total catastrophe unless we have far more faith in our fellow human beings than we do right now. Within a single generation we need to form not only a planetary government, but a world society that is more organically integrated than any known historical culture — a society wherein each person trusts her fellow citizen more than she now trusts her husband, sister, or father. This is not some audaciously speculative, utopian proposal. By the time you arrive at Chapter 6 you should realize why it is a survival imperative.

Unfortunately, the emergence of this world state deeply unified by a single ethos, is going to be a violent process. That is not to say it will come about through arbitrary force. Rather, quite in conformity with Schmitt's analysis of the true constitution of a state, a distinct world-historical folk will undergo an alchemical transformation in the cauldron of the technological apocalypse and reclaim its destiny of planetary hegemony. In the face of external and internal enemies incapable of responsibly shepherding the Earth through the technological apocalypse, the Indo-European community will define itself as a sovereign nation of planetary or even interplanetary scope. With their common origin in a single ethnicity and language, the Indo-Europeans spread out from a proto-Iranian or Aryan heartland in Ukraine (Scythia) and the Caucasus to form the European, Iranian, Hindu, and Buddhist civilizations. They continued to influence each other through the quintessentially Indo-European nation, *Irân-Shahr* or the Aryan Imperium (known in the West as "the Persian Empire"). Collectively, their contributions to world religions, the arts, philosophy, science, and technology, establish the Indo-Europeans as the most viable contenders to form a world state from out of the planetary emergency.

Chapters 6 and 7 focus on the history and culture of Indo-Europeans, and argue that our convergence into a single civilization of planetary scope can become the basis for the world state that we so badly need to survive the technological apocalypse. The alternative is Islam or China stepping in to save Humanity from the Promethean genius of the Indo-Europeans. Instead, we should harness this Faustian spirit in a way that will avert inhuman monstrosities and lead us into a future of superhuman flourishing. Since, as Samuel Huntington understands, Islam is playing the most catalytic role in the clash of civilizations, from out of the entire Indo-European community Iran is the nation whose role in the Third World War will be most decisive for our common future. In addition to being the cross-cultural nexus of the Indo-European world for more than 3,000 years, *Irân-Shahr* or "the Aryan Imperium" offers the West principles and values that have already deeply influenced its own and that can catalyze a cultural revitalization beyond the failure of modern concepts such as liberal democracy and universal human rights.

In the seventh and final chapter, I discuss how since 2012 a Neo-Zoroastrian revival known as the Iranian Renaissance has been spreading in Greater Iran like a firestorm. It is profoundly significant in that the Persians and Kurds are the first of the Aryan peoples to have gone through the entire Abrahamic religious tradition and come out the other side. Europe is only now on the verge of being conquered by Islam, whereas the Iranians are finally overcoming 1,400 years of oppressive Arab-Islamic colonization that parasitically misappropriated the Persian genius in the arts and sciences. Sharing a vast border with China in Iranian Central Asia or Khorasan, and being situated in the heart of an emerging Islamic Caliphate controlled by ISIL in its Western territories and Al-Qaeda to the East, a renaissance of Greater Iran or *Irân-Shahr* will be the spearhead of the war for an Indo-European World Order.

If we are able to someday go out and explore the cosmos, if we make it past this technological apocalypse, we will find that the world state of emergency has occurred in the histories of all intelligent species. While technological developments may continue beyond this point, it really does represent a kind of singularity. The period before it and the period after it will be incommensurate with one another. In that sense it is a classical catastrophe. A one shot deal. There comes a time when technologies are developed that call into question the very form of life of a certain species, and the members of that species have to collectively grapple with coming to a self-understanding that allows them to use these technologies for the purpose of flourishing rather than a degradation of their existence. Ready or not, the time has come to ask, "Who speaks for Earth?"

CHAPTER 1

The Third World War

On September 11, 2001 we entered the Third World War. Although it has rarely been recognized as such, that is ironically because it is the first true world war. Unlike, World War I and World War II, which were international conflicts, the present global war is not a confrontation between nation states that just happens to extend over most of the planet. Rather, it is a war of cultural worlds — a clash of civilizations. Consequently, it will also be far more convoluted and protracted, especially insofar as the policy making elite of the West in particular continues to deny the nature of this war, despite the fact that it was so predictable that its dynamics should have been grasped even before it began. In fact, it *was* predicted. In his 1996 book entitled *The Clash of Civilizations and the Remaking of World Order*, the Harvard political scientist Samuel P. Huntington analyzes the decline of the West, with its International System, and warns of an impending civilizational war on a planetary scale. That war began five years later, one beautiful late summer morning in lower Manhattan. I remember it well. I was there.

In *The Clash of Civilizations*, Huntington contends that the idea that Western liberal democracy would be universally accepted simply because it triumphed over Soviet communism is a "single alternative fallacy."[1] In a post Cold War world, there are as many major alternatives to the West as there are non-Western civilizations. Furthermore, the idea that

---

[1] Samuel P. Huntington, *The Clash of Civilizations and the Remaking of World Order* (New York: Simon and Schuster, 2003), 66.

global connectivity through international trade will somehow foster a universal civilization is also gravely mistaken. Huntington points out that international trade was a record high in 1913, just before the major countries of the world decided to tear each other to pieces in the First World War. It might also be relevant to point out that anyone who thinks China's MFN trading status with the United States or its extensive dealings with Walmart would prevent a Third World War to challenge the rise of China forgets that Ford and IBM continued to do business with Nazi Germany well into the Second World War. In fact, as Huntington sees it, the depersonalizing forces of international trade actually intensify the need for alienated participants in it to identify themselves with their own civilization. A Frenchman and a German meeting somewhere in Europe may be just that, but in rough oil negotiations with Saudi, Iraqi, and Qatari members of OPEC they become two Europeans dealing with a bunch of Arab Muslims.[2] This is becoming especially clear with respect to Chinese-dominated Asia and the West.

At the 1993 Vienna Human Rights Conference a United States delegation led by Secretary of State Warren Christopher denounced "cultural relativism" in the face of Islamic and Confucian rejections of "Western universalism."[3] American efforts to use economically punitive measures to enforce compliance with human rights standards in China (post Tiananmen square) and the Islamic world have been a failure.[4] Beginning with the Vienna conference in 1993, Asian and Islamic countries pointed out that when the Universal Declaration of Human Rights was written Western countries were responsible for half the world's economic output, and asserted that the decline in Western economic power somehow translated into the irrelevancy of the declaration (as an attempt to assert Western values as universal values).[5] As far as promotion of democracy is concerned, Huntington rightly points out that it need not lead to

---

[2] Ibid., 67.

[3] Ibid., 38.

[4] Ibid., 194–195.

[5] Ibid., 196.

Westernization.⁶ Events in the Islamic World since the so-called 'Arab Spring' have certainly validated this theory. In the Arab world post 9/11, democratization has in each case led to theocratic Islamization.

According to Huntington, Marxism and every other significant political ideology — including liberalism, anarchism, corporatism, social democracy, conservatism, nationalism, and fascism — are products of Western civilization. By contrast, he claims that the West has never produced a major religion and its 'own' religions are imports from non-Western cultures. This is very significant insofar as the era of political ideology is now over and the clash of civilizations that is to replace it will be primarily driven by religious cultures.⁷

By 2025, Islam will replace Christianity as the world's dominant religion by population.⁸ For Huntington this is especially significant in that he sees religious proselytism and conversion as the primary means of the spread of a civilization, not the highbrow study of the philosophical, political or economic ideologies espoused by Western intellectual authors. Islam has an edge over Christianity, because in addition to a higher population growth rate in Muslim lands, unlike purely confessional Christianity, once conversion to Islam has occurred on penalty of death all of one's descendants are also Muslim and cannot renounce the religion of their forefathers without risking death.

The Soviet war in Afghanistan from 1979–1989 was a kind of prelude to the outbreak of the civilizational world war in 2001. Although it began as an attempt to shore up the regime of a satellite state, the USSR's war in Afghanistan ended up being the first modern Islamic jihad. The success of this *jihad*, as compared to the relative failure of secular and nationalist approaches such as those taken by Nasser or the Shah of Iran, gave a tremendous boost of self-confidence to Islamic Civilization. It was the transitional point between the ideological world order of the Cold War

---

6   Ibid., 193, 198.

7   Ibid., 53–54.

8   Ibid., 65–66.

and the new clash of civilizations.[9] Consequently, neither the Americans nor the Soviet Union — who were thinking in Cold War terms — understood what was at stake. The first modern jihad could not have been accomplished without nearly a billion dollars of American funding — nearly $500 million directly, and another half a billion through Saudi Arabia (the majority of whose wealth derives from oil exports to the US). Furthermore, the US gave operational support to Pakistani intelligence to organize Islamist guerrilla fighters at camps in Pakistan, and armed them with sophisticated weapons such as Stinger missiles with which to fight the Soviets in Afghanistan.[10] Overlooking this US aid, the narrative the Islamists wove out of *their* victory over "godless communists" was that if they could beat one superpower, there was nothing to say they could not beat the other.

There was a time, in the late 19th and early 20th centuries, when autocratic leaders of non-Western societies believed that the path to modernization needed to be a path of Westernization. Huntington identifies this position above all with Mustafa Kamal Ataturk, who violently transformed the Islamic Caliphate of the Ottoman Empire into the modern Western-style secular republic of Turkey. Huntington argues that this is only the first, naive stage of a process that ultimately leads to the reaffirmation of traditional non-Western cultures.[11] These cultures superficially westernize in order to modernize, but as technical modernization leads to a high level of economic development and increased military capability, it also leads to an identity crisis of alienation and social anomie. This tension is resolved by rejecting *secular* Western cultural influence in favor of a reaffirmation of the traditional native culture, especially its *religious basis*, and attributing the successes of economic and military advancement to the superiority of their own civilizational values to those of the West. That has now happened in Turkey as well.

This perverse inversion has taken place above all in post-Maoist China. Interestingly, Huntington sees the triumph of Marxism in Russia

---

9   Ibid., 246–247.

10 Ibid., 247.

11 Ibid., 74–77.

and then in the Chinese sphere (China, North Korea, Vietnam) as the transition from the European International System to a post-European multi-civilizational system.[12] This is quite bizarre since of all European ideologies, Marxism is the most committed to the convergence of all distinct traditional cultures into a single cosmopolitan civilization valorizing rational ideals uniquely developed (or as they would have it "discovered") by the European philosophical tradition from Plato through to Descartes, Kant, and Hegel. Huntington addresses this seeming paradox by explaining that modernizing elites within Orthodox and Asian civilization imported Marxism, but that it was used primarily as a means to eliminate obstacles to 'modernization' construed as technical progress. The deeper philosophical roots of Marxist ideology were not absorbed by any significant segment of these societies. Instead, if anything, a superficial understanding of Marxism was used to mobilize these societies against Western colonialism. Ultimately, in China, this evolved into a very un-Marxist nationalist affirmation of Chinese ethnic superiority to the West — an affirmation now possible on account of successful technical 'modernization' without attendant cultural Westernization that would have provided the deep civilizational context for Marx's own universalistic cosmopolitanism.

Three out of the four Asian "tigers" are Chinese (only South Korea is not).[13] Although ethnic Chinese are a minority in Southeast Asian countries like Indonesia and Malaysia, the vast majority of the capital that accounts for the recent economic productivity of those countries is concentrated in their hands. China has spread its economic influence in East and Southeast Asia through *quanxi* connections of language, culture, and frankly *racist* proclivities. As Lee Kuan Yew put it: "People feel a natural empathy for those who share their physical attributes… It makes for easy rapport and trust, which is the foundation for all business relations."[14] 80% of direct foreign investment in China comes from overseas Chinese, whereas under 7% comes from Japan and under 5% from the

---

12   Ibid., 52.

13   Ibid., 169.

14   Ibid., 170.

United States.¹⁵ Absorption of the several other Chinas has given the PRC on the mainland every material asset of the West, in addition to the land resources and vast labor pool that it already had. This includes the world-class entrepreneurial acumen of Hong Kong, the global communications network and capital investment of Singapore, and the high technology manufacturing capabilities of Taiwan.¹⁶

The expansion of Chinese migrants and business interests into Siberia has been viewed by some Russian military officials as "a peaceful conquest of the Russian Far East."¹⁷ The Chinese might ultimately wish to reclaim Mongolia, which was detached from China by the Russians at the close of World War I. Chinese influence in the former Soviet republics of Central Asia has also greatly intensified. Huntington writes: "Russia and China united would decisively tilt the Eurasian balance against the West and arouse all the concerns that existed about the Sino-Soviet relationship in the 1950s."¹⁸ Events subsequent to his publication have born out this scenario, with the rise of the *Shanghai Cooperation Organization*.

The United States has thus far been willing to have only a quasi-world empire, leading a world of sovereign nation states mostly by example, as the single most powerful state among them. This has, however, been predicated on America's ability to forestall the rise of some other dominant power in Europe or Asia. Two world wars were fought against Germany to prevent it from becoming the dominant power in Europe. The second of these world wars also involved a devastating assault on Japan to prevent its imperial dominance of the Asia-Pacific region. Finally, the cold war against the Communist states of the Soviet Union and the People's Republic of China were also aimed at containing their ambitions for global dominance in Europe and Asia. Whereas Westerners, Americans included, tend to think in terms of maintaining a favorable position within the context of a strategic balance of powers, Asians affirm hierarchical and

---

15   Ibid., 170–171.

16   Ibid., 171.

17   Ibid., 243.

18   Ibid., 241.

centralized authority. If post-communist China — which Lee Kuan Yew has referred to as not "just another big player" entering the "great game" but "the biggest player in the history of man" — were to rise to a position of dominance in Asia, and potentially make a bid for global dominance, the US would either have to yield or radically change its approach.[19]

The values of China's Confucian civilization consist of "authority, hierarchy, the subordination of individual rights and interests, the importance of consensus, the avoidance of confrontation, 'saving face', and, in general, the supremacy of the state over society and of society over the individual."[20] He also claims that while Americans tend to think in terms of maximizing immediate gains, Asians have an almost millennial perspective. In a terser definition of "China's Confucian heritage", Huntington highlights "its emphasis on authority, order, hierarchy, and the supremacy of the collectivity over the individual."[21] The Confucian world-view has no room for "two suns in the sky" and the "Middle Kingdom" is not a vision of China compatible with a multi-polar world.[22] Whereas Westerners have long understood the dynamic re-balancing of powers as *generative* strife, "peace and hegemony" are one in the traditionalist Asian mind.[23] Huntington argues that as the West is displaced by Chinese Civilization English will be supplanted by Mandarin as the world's *lingua franca*.[24] That transition has already occurred in East and Southeast Asia.

Yet Huntington misunderstands the ambitions of the People's Republic of China as being restricted to the following: "to become the champion of Chinese culture, the core state civilizational magnet toward which all other Chinese communities would orient themselves, and to resume its historical position, which it lost in the nineteenth century, as

---

19 Ibid., 229–231.
20 Ibid., 225.
21 Ibid., 238.
22 Ibid., 234.
23 Ibid., 238.
24 Ibid., 63.

the hegemonic power in East Asia."[25] The Chinese government's advocacy of Chinese dominance is primarily racial, not cultural. The idea is not to spread Chinese culture around the world in such a way that anyone who adopts it is considered Chinese, just as anyone who adopts and *adapts* Western culture is considered a Westerner regardless of race. Rather, it believes in the *mirror test* — if you want to know whether you are Chinese that is ultimately decided by looking into the mirror.[26] Even within the limited territory of the PRC, let alone the emerging Shanghai Cooperation Organization, promotion of Chinese identity has meant the Han ethnicity's domination of all others (Tibetans, Mongols, Uighurs, etc.). With its combination of an insular racism that conflates culture with blood and its economic corporatism, post-Mao China is at this point really a *fascist* government with imperial ambitions of potentially global scope.

Huntington equates universal civilization with universal power. The only way a particular civilization can become universal, in his view, is for it to impose itself by force throughout the world — as he believes the Romans did within the compass of the Classical world.[27] He does not buy that culture and ideology can seduce people into adopting another civilization as their own. What makes these attractive is their association with material wealth and military prowess. Consequently, the decline of the West in both these regards is precipitating the decline in the acceptance of Western values — such as "human rights" — as universal values.[28] He quotes an Asian head of state in a perverse statement that addressed European government officials at a 1996 meeting, to the effect that: "Asian values are universal values. European values are European values."[29]

In this connection Huntington makes an interesting remark about what he sees as an internal contradiction in the position of most people who advocate Western Civilization as a universal civilization:

---

25 Ibid., 168.

26 Ibid., 169.

27 Ibid., 91.

28 Ibid., 92.

29 Ibid., 109.

> The link between power and culture is almost universally ignored by those who argue that a universal civilization is and should be emerging as well as by those who argue that Westernization is a prerequisite to modernization. They refuse to recognize that the logic of their argument requires them to support the expansion and consolidation of Western domination of the world, and that if other societies are left free to shape their own destinies they reinvigorate old creeds, habits, practices which, according to the universalists, are inimical to progress. The people who argue the virtues of a universal civilization, however, do not usually argue the virtues of a universal empire.[30]

When the West was strongest, in the late colonial era, non-Western societies challenged the West by claiming the universality of Western values — in effect, by accusing the West of hypocrisy and making a bid to join Western Civilization on their own terms as equal partners. Now, when the West is perceived as increasingly weakened by self-doubt and in a state of terminal decline, increasingly affluent and militarily capable non-Western societies are challenging the West on their own terms — by rejecting "decadent" Western values altogether and promoting their own culture as evidently superior and responsible for their success.[31] Huntington refers to this transition as a "process of indigenization" and shows how it can even happen within a single generation. He offers two compelling examples. One is the transformation of the Anglophile Christian Cambridge graduate, Harry Lee, into the Mandarin-speaking Neo-Confucian pro-China strongman of Singapore, Lee Kuan Yew. The other is the similar transformation of the English lawyer M.A. Jinnah, an secular intellectual and Oxford graduate, into Quaid-i-Azam — the fervently religious apostolic founder of the Islamic Republic of Pakistan.[32]

The indigenization process is exacerbated by what Huntington calls "the democracy paradox" — although I do not see anything paradoxical about it. He rightly points out that the adoption of the Western political institution of democracy by non-Western countries has actually led to the overthrow of Westernized elites in those countries and the empowerment

---

30 Ibid., 92.

31 Ibid., 93.

32 Ibid., 93–94.

of people promoting the native culture and religion. Non-western democracies are "ethnic, nationalist, and religious in character" because: "Democratization conflicts with Westernization, and democracy is inherently a parochializing not a cosmopolitanizing process."[33] This equation of Westernization with "a cosmopolitanizing process" and the recognition that it is actually undermined by the export of democracy to non-Western societies is an astute observation on Huntington's part, but one from which he does not draw the proper conclusion.

The most remarkable element of the indigenization attendant to democratization of non-Western societies is the revival of religion. In the early 20th century, it was believed both by intellectuals in the West and powerful Westernized elites outside it, that religion was on the decline and that, on account of advances in scientific knowledge and more widespread education and economic development, in the future society would be organized on a secular humanistic, rational — or at least pragmatic basis.[34] In 1989, just before the collapse of the Soviet Union, there were only 160 functioning mosques and only one madrassah in Central Asia. Within five years, just after the disintegration of the Soviet Union, there were 10,000 mosques and ten madrassahs.[35] The attempt to eliminate "a sacred foundation for the organization of society" failed miserably, and Huntington believes that this is because: "People do not live by reason alone."[36] As the Confucian leader of Singapore explains, it is impossible for a society to do without "a quest for some higher explanations about man's purpose, about why we are here."[37] Religions not only provide purposeful direction to people's lives, they also provide the most fundamental basis for the human need to define oneself with respect to an in-group and various out-groups.[38]

---

33 Ibid., 94.

34 Ibid., 95.

35 Ibid., 96.

36 Ibid., 96–97.

37 Ibid., 97.

38 Ibid., 97.

Advocates of "multiculturalism" in the United States have weakened its social fabric by calling into question its identification with Western Civilization, arguing against teaching the Western literary canon because they see the West only in terms of its 'sinful' colonialist 'crimes', without advocating any clear alternative other than ethnic balkanization and a ghettoized North America.[39] They have undermined the fundamental ideal of the United States, protection of the rights of *the individual*, by advocating instead an identity politics of groups defined in terms of race, ethnicity, sex, and sexual orientation.[40] Multiculturalists can only damage or destroy the American cultural relationship with its European civilizational cradle, but they can offer nothing to replace it.[41] Of course, their identity politics can also be appropriated for a reactionary reaffirmation of the foundational European identity of North America.

There is nothing to suggest that the pattern of civilizational rise and decline is any different for the West than it has been for other civilizations.[42] Huntington adopts Carol Quigley's understanding of the decline of civilizations as marked by a point when: 1) people live decadently off their capital instead of investing their surplus value in expansive ends or bold new ways of doing things (the Manhattan Project, the Apollo Space Program, ARPANET, etc.); 2) social disintegration sets in, as is evident in civil strife and the rise of new religious cults; 3) a civilization is no longer willing to defend itself against encroaching 'barbarians' from a more youthful and vigorous civilization.[43] The West, especially the United States, is showing signs of all these developments. Lower economic growth rates, savings rates, investment rates, increases in crime, drug use, violence, family decay, declining interpersonal trust, the replacement of a good

---

39 Ibid., 305.
40 Ibid., 306.
41 Ibid., 307.
42 Ibid., 302.
43 Ibid., 303.

work ethic with a cult of personal indulgence, a decreasing concern with scholastic achievement, and lower military force and spending ratios.[44]

Mass immigration and migration has tremendous strategic significance in the clash of civilizations.[45] The demographic invasion of the rest of the world by tens of millions of Europeans, from 1821 to 1924, was one decisive factor in establishing the dominance of Western Civilization over decimated or colonized native populations. What is significant to the thesis of civilizational clash is that, this tide has now reversed. Beginning in the 1960s and 70s, the United States and Europe changed immigration laws with quotas favoring Europeans and economic and social changes in Asian and Islamic countries also led to an increasing percentage of those populations prone to immigration. Consequently, by the 1990s, major European cities, such as Paris and London, and key regions of the United States, such as California and New York, were experiencing a major demographic shift involving immigrants from non-Western civilizations — especially Asia and Islam.[46] In France, where Africans that are culturally/linguistically French are accepted, the fear of this invasion is civilizational more than racial.[47] Demographic shifts matter. Huntington suggests that it was the demographic shift in Lebanon in favor of Shiite Muslims over Francophile Maronite Christians that made the constitutional order there disintegrate in the 1970s and ultimately led to protracted civil war, out of which emerged a Hezbollah client state of Khomeini's Iran.[48] He sees a similar dynamic at work in 1980s Chechnya.[49]

Huntington believes that the most intense civilizational clashes will be those between Islam and the West and China and the West, with Japan, Russia, and India acting as "swing" civilizations that may align with the

---

44   Ibid., 304.

45   Ibid., 198–199.

46   Ibid., 199.

47   Ibid., 200.

48   Ibid., 259.

49   Ibid., 260.

West or its antagonists.[50] However, he thinks that profound cultural differences will limit a potential "Confucian-Islamic" alliance to the kind of working relationship against a common enemy that the Allies and Stalin developed in their struggle against Hitler's Fascist Axis. China and Pakistan have developed a very tight strategic alliance since the 1970s. The Chinese have provided Pakistan with all manner of military equipment, including long range missiles and even assistance with the development of their nuclear weapons. In exchange the Pakistanis have transferred US midair refueling technology and Stinger missiles to Pakistan.[51]

Huntington claims to show, through a statistical analysis, that although at the time of his writing Muslims accounted for only one-fifth of the world population, they have been involved in the vast majority of conflicts in Northern Africa, Southeastern Europe, the Caucasus, the Middle East, Central and Southeast Asia.[52] They are the only common denominator in hundreds of conflicts against a diversity of other unrelated ethnic groups spanning three continents. They are everybody else's trouble. There is also more intracivilizational conflict among Muslims than within any other civilization.[53] The average force ratios and military effort ratios (the percentage of the population in the military and the percentage of the budget spent on the military, respectively) of Islamic countries is roughly twice that of Western nations.[54] These facts taken together suggest that Islam is in a new state of expansion, similar to that of the colonial West — which was also constantly at war with itself until it had successfully conquered most of the planet at the end of World War I.

In 1800, through their colonial empires Europeans controlled 35 percent of the earth's land surface. This increased to 67 percent in 1878 and peaked at 84 percent in 1914 — the year that the World War began. In 1900 about 30 percent of the world's population were Westerners, and by 1920

---

50   Ibid., 185.
51   Ibid., 189.
52   Ibid., 256–257.
53   Ibid., 257–258.
54   Ibid., 258.

Western colonial government's ruled over 48 percent of the world's population. When Woodrow Wilson, Lloyd George, and Georges Clemenceau conferred at the Paris Conference in 1919, between them they controlled the entire world — carving up their colonial realms and lands conquered in World War I into artificial countries of their own design. They even had the capacity to extract economic concessions from China and to militarily intervene in Russia, which was in a state of chaos only a few years into the Revolution in part stirred up by wartime privation.[55] The League of Nations that emerged from out of that conference could really have become a world government that established Western Civilization as Earth's universal civilization.

By the end of the 20th century, Westerners made up no more than 13 percent of humanity, putting them in fourth place behind the Chinese, Islamic, and Hindu civilizations, and Western governments no longer ruled over any significant populations other than their own.[56] The industrial manufacturing output of Western nations peaked in 1928 at 84.2 percent of world manufacturing, whereas that figure has now dropped to a mere 30 percent. International trade — which basically meant trade between European colonial entities, the free colonies of North America included — was more intensive in 1910 than at any other time until it peaked in the 1980s. In that era, "Civilization meant Western civilization" and the International Law that governed it was the Western European legal tradition established by Grotius and set in place at the Treaty of Westphalia.[57] This was replaced with a delusional attempt to end war, one that disregarded the ontological foundation of the very concept of political order.

At the core of this new so-called International System was the *Universal Declaration of Human Rights* (UDHR) drafted in 1948, as well as an insistence on the globalization of democracy by the United Nations Organization and related institutions. The 1948 *Universal Declaration of*

---

55 Ibid., 91.

56 Ibid., 84.

57 Ibid., 52.

Human Rights (UDHR) protects freedom of religion unconditionally, and declares that popular sovereignty is the basis of the legal authority of governments. The drafters formulated these rights in an unqualified manner that renders the Declaration incoherent and self-vitiating, because it inadvertently allows the 'moral majority' of a religious democracy to deny dissenting minorities the very human rights that are protected by the Declaration. Certain of the drafters of the declaration were aware of this possibility, but lost their debates to properly address it.

In his seminal *Letter Concerning Toleration*, the most stringent limit that Locke puts on the liberty of rites of worship and articles of faith is that no religion will be tolerated if it entails a system of political sovereignty with the claim of the sole authority to make laws binding on all people:

> These therefore, and the like, who attribute unto the Faithful, Religious and Orthodox, that is, in plain terms, unto themselves, any peculiar Privilege of Power above other Mortals, in Civil Concernments; or who, upon pretense of Religion, do challenge any manner of Authority over such, as are not associated with them in the Ecclesiastical Communion; I say these have no right to be tolerated by the Magistrate; as neither those that will not own and teach the Duty of tolerating All men in matters of meer Religion. For what do these and the like Doctrines signifie, but that they may, and are ready upon any occasion to seise the Government, and possess themselves of the Estates and Fortunes of their Fellow-Subjects; and that they only ask leave to be tolerated by the Magistrate so long until they find themselves strong enough to effect it? ... For by this means [by tolerating such a religion] the Magistrate would give way to the settling of a foreign Jurisdiction in his own Country, and suffer his own People to be listed, as it were, for Souldiers against his own Government.[58]

This is perhaps the most important passage in all writings on religious toleration by anyone. The most basic principle of religious toleration is that a tolerated religion can have no legal implications and its tenants can have no claim to sovereign authority or inherently challenge the sovereign authority of the state as the sole arbiter of laws affecting worldly Life, Liberty and Property. If this principle should therefore prove false, religious toleration as such falls together with it. Locke acknowledges that

---

58 John Locke, *A Letter Concerning Toleration* (Indianapolis: Hackett, 1983), 50.

Israelite Judaism *as established by God himself* is an example of an inherently political religion. He even goes so far as to admit that, for Jews, the political aspects of their religion are *not optional* and that they *do* qualify as "rites of worship" and "articles of faith" as he defines them.[59] Locke claims that this revealed positive Law of Moses was restricted to the Jews alone, *and that since that time God has never again been the lawgiver of any other human society*. According to Locke, not only was the mission of Jesus Christ not that of a positive lawgiver on behalf of God, but this Christian mission even *abolished the Mosaic right to the use of force for the implementation of divine laws.*

It is on the basis of this last claim that Locke argues that anyone who is commanded by the magistrate to do something against their faith may conscientiously object, but they must peacefully submit to imprisonment or any other punishment the magistrate sees fit.[60] According to Locke, if such a man is right he will be vindicated by God in the afterlife, and so his conscientious objection in this world should not disturb the public peace by challenging the sovereignty of the magistrate. Locke admits that his argument for toleration would not hold if it were shown that "there is a Commonwealth, at this time" wherein "Laws established there concerning the Worship of One Invisible Deity, [are] the Civil Laws of that People, and a part of their Political Government; in which God himself [is] the Legislator."[61] In fact, Islam *is* a contemporary socio-religious-political phenomenon that arose *after* the Christian example that Locke takes to have negated such a Mosaic-style Religion.

While the historically conditioned incoherence of Judeo-Christianity militates against fundamentalist interpretations of the *New Testament*, a single careful and honest reading of the *Quran* shows that the terms 'Islamic fundamentalism' and 'political Islam' are redundancies. There is an extensive interpretive Islamic legal tradition, with five main schools, but the *Quran* itself contains a clear body of divinely mandated civil law,

---

59   Ibid., 44–45.

60   Ibid., 50.

61   Ibid., 45.

established by a single legislator, namely Muhammad, whose reality as a (relatively recent) historical figure is well established. All of the schools of Islamic law agree on these Quranic laws, and while these laws are not comprehensive in the absence of the Islamic legal tradition, the tradition only elaborates these laws and extrapolates from them. As Wael B. Hallaq explains in his study of *The Origins and Evolution of Islamic Law*, Quranic law as established by Muhammad during his own lifetime is the basis of all subsequent Islamic law as elaborated by the jurists of the five schools.[62] These jurists tried to make Islamic law or *sharia* comprehensive in various ways by inference from which existing Arabian laws Muhammad *did not modify or repeal* by means of verses from the *Quran*, during his tenure as governor of Medina.[63]

The *Quran* contains some 500 legal verses, with the length of these legal verses being, on average two or three times that of non-legal verses.[64] Consequently, in his study of the origins of Islamic Law, Hallaq draws the conclusion that "the *Quran* contains no less legal material than does the *Torah*, which is commonly known as 'the Law.'" However, the *Quran* was composed in only 22 years (610–632), by a single man. It was written down in fragments, on rocks and on the shoulder blade bones of camels, during his own lifetime by scribes, and also memorized by professional bards or *ha'afiz*.[65] The text, *as we have it today*, was compiled in manuscript form *by Muhammad's own scribe*, Zayd bin Thabit, under the direction of the Caliph Uthman, around 650 — a mere 20 years after Muhammad's death.[66] Compare this to the *New Testament*, compiled from out of *tens* of radically divergent texts, by as many authors with essentially different motives, *some 300 years* after the life of Jesus.[67] These are all factors that make the *Quran* a much more coherent, concise, and authoritative

---

62 Wael B. Hallaq, *The Origins and Evolution of Islamic Law* (New York: Cambridge University Press, 2005), 8–28.

63 Ibid., 19–25.

64 Ibid., 21.

65 Ibid., 33.

66 Ibid.

67 Burton L. Mack, *Who Wrote the New Testament* (New York: Harper Collins, 1995).

revealed religious text for careful comparison with the relevant articles of the *Universal Declaration of Human Rights* (UDHR) to the end of demonstrating fundamental problems with the ideology of human rights as such.

Saudi Arabia's stated grounds for its abstention in the vote to ratify the UDHR seems an appropriate place to begin an examination of whether Islam inherently contradicts the principles of universal human rights. After having led many other muslim nations in voting unsuccessfully against the present wording of certain of the UDHR's articles, the Saudis refused to ratify the Declaration. The Saudi delegation explicitly cited opposition to Article 16's provision for equal marriage rights and Article 18's right "to change his religion or belief" as reasons for their abstention from ratification of the *Universal Declaration of Human Rights*.

Regarding Article 16, the Saudi delegation wanted it to simply read "Men and women of legal matrimonial age within every country have the right to marry and found a family," *without* the following clauses presently part of the article: "They are entitled to equal rights as to marriage, during marriage, and at its dissolution. Marriage shall be entered into only with the free and full consent of the intending spouses."[68] The equal marriage rights protected by Article 16 reflect the UDHR's explicit and repeated rejection of sex as a basis for abrogating the universality of human rights and its assertion that any legal system that institutionalizes sexual discrimination is illegitimate. Article 2 of the UDHR states: "Everyone is entitled to all the rights and freedoms set forth in this Declaration, without distinction of any kind, such as race, colour, *sex*..." Article 7 of the UDHR reads: "All are equal before the law and are entitled without any discrimination to equal protection of the law. All are entitled to equal protection against any discrimination in violation of this Declaration and against any incitement to such discrimination." Representative Jamil Baroody of Saudi Arabia seemed to sum-up his delegation's objections to women's equal rights when he accused the drafters of the Declaration of having

---

68 Johannes Morsink, *The Universal Declaration of Human Rights: Origins, Drafting, and Intent* (Philadelphia: University of Pennsylvania Press, 1999), 24.

> ...for the most part taken into consideration only the standards recognized by Western civilization and had ignored more ancient civilizations which were past the experimental stage, and the institutions of which, for example marriage, had proved their wisdom through the centuries. It was not for the Committee to proclaim the superiority of one civilization over all the others or to establish uniform standards for all the countries in the world.[69]

The nature of this protest evinces a fundamental Saudi opposition to the idea of universal human rights — which would be by definition "uniform standards for all the countries in the world", granted that these countries are all populated by human individuals. Furthermore, Baroody was clearly referring to Islamic civilization in his suggestion that the allegedly 'Western' standards informing the Declaration are representative of a civilization still at an "experimental stage", while the law of certain ancient civilizations had "proved its wisdom through the centuries". His broad use of the term "ancient civilizations" was shrewdly pitched to appeal to brahmanical Hindus who, despite being 'idolatrous enemies of Islam' from the Saudi perspective, also view history as degeneration from a golden age rather than as progress.

This radically traditionalist view of the Saudi delegation expresses an absolutely closed-minded attitude that is antithetical to the progressive spirit that has been intrinsic to the modern idea of human rights since its conception in the minds of Enlightenment philosophers. Ambassador Baroody's objection to equal human rights for women is, however, authentically Islamic. Verse 4:34 of the *Quran* reads:

> Men have authority over women because God has made the one superior to the other, and because they spend their wealth to maintain them. Good women are obedient. They guard their unseen parts because God has guarded them. As for those from whom you fear disobedience, admonish them and send them to beds apart and beat them. Then if they obey you, take no further action against them. Surely, God is high, supreme.

Women constitute half of the human race, and so the question of whether Islam could in any way be compatible with equal human rights for women

---

69 Ibid.

deserves a more in depth treatment than any other point of comparison between the UDHR and the *Quran*. A very strong case can be made that Allah views the subordination of women to men as Just and Natural. There are many verses other than 4:34 that suggest this. In 4:11 a woman is given the right to only half of the inheritance entitled to a man. In 2:223 men are told that they may sleep with their women whenever it pleases *them* to do so. In 43:15–18 and 53:27 the notion of female divinity is ridiculed and in the same breath the idea that male heirs are more desirable than female children is affirmed. Verses 78:31–33; 55:54–66; 56:35–38; 52:19–20; and 37:40 all objectify women as sexual playthings for men in paradise, while there is never any mention in the *Quran* of heavenly sexual consorts for women. Verse 2:282 requires two female witnesses to compensate for the lack of only one of two prescribed male witnesses at a legal proceeding, on account of the feeblemindedness of women, in a clear violation of Article 7 of the UDHR on *equality before the law*, as well as in contradiction to the claim of Article 1 that "All human beings are born free and equal... endowed with reason and conscience":

> If the debtor be an ignorant or feeble-minded person, or one who cannot dictate, let his guardian dictate for him in fairness. Call in two male witnesses from among you, but if two men cannot be found, then one man and two women whom you judge fit to act as witnesses; so that if either of them commit an error, the other will remind her.[70] (*Quran* 2:282)

Verses 23:1–6 in the *Quran* allow a man to have sexual relations with as many slave women as he has seized in battle (in addition to his legal wives), whereas a woman is the sole sexual possession of her lawful husband. Verse 2:222 burdens women with the stigma of being ritually unclean during their monthly menstrual period, which, given Islam's code of ritual purity, prevents them from religious leadership that is indistinguishable from the highest levels of political power. Verse 4:16 enjoins men to confine women convicted of adultery to their houses until death overtakes them, but the same verse says that adulterous men (which interestingly is only homosexually conceived) should be let alone if they repent after

---

70   N.J. Dawood (translator), *The Koran* (New York: Penguin Classics, 1995).

a corrective punishment. This is clearly a violation of the *equality before the law* mandated by Article 7, not to mention the prohibition of cruel punishment in Article 5.

Furthermore, certain Quranic verses discriminating against women, such as beatings for fear of disobedience or life imprisonment within the home as a punishment for adultery, are so harsh that they raise the question of cruel and inhumane treatment or punishment. Article 5 of the *Universal Declaration of Human Rights* demands that: "No one shall be subjected to torture or to cruel, inhuman or degrading treatment or punishment." The *Quran*'s code of crime and punishment presents us with some striking violations of this article.

Cutting off arms and legs on alternate sides is permitted in the name of a just cause (25:68) such as the fight to the death against unbelievers: "Those that make war against God and His apostle and spread disorder in the land shall be put to death or crucified or have their hands and feet cut off on alternate sides, or be banished from the land…Fight valiantly for His cause, so that you may triumph." (5:33–35) Decapitation is also prescribed as a method of punishing unbelievers in the context of a war against apostasy: "Thus God lays down for mankind their rules of conduct. When you meet the unbelievers in the battlefield strike off their heads and, when you have laid them low, bind your captives firmly." (47:3–4) In verse 8:12 the call for beheading is accompanied by a more sadistic punishment, the severing of fingertips: "I shall be with you. Give courage to the believers. I shall cast terror into the hearts of the infidels. Strike off their heads, strike off the very tips of their fingers!"

We also see cruel corporal punishment in the *Quran*'s penalty for theft, which is amputation: "As for the man or woman who is guilty of theft, cut off their hands to punish them for their crimes. That is the punishment enjoined by God. God is mighty and wise." (5:38) As noted above, Verse 4:16 enjoins men to confine women convicted of adultery to their houses until death overtakes them, unless by some grace of God their husband is killed or their house burns down and they are able to get loose: "If any of your women commit fornication, call in four witnesses from among yourselves against them; if they testify to their guilt confine them to their houses till death overtakes them or till God finds another way for them."

The UDHR recognizes slavery as a particularly offensive form of "inhuman or degrading treatment" so that a condemnation of slavery merits its own article, namely Article 4: "No one shall be held in slavery or servitude; slavery and the slave trade shall be prohibited in all their forms." By contrast, the approval of slavery may be seen in a number of Quranic verses. One of these passages concerns a man's right to have sexual relations with as many slave girls as he has seized "by [his] right hand", i.e. in battle:

> Blessed are the believers, who are humble in their prayers; who avoid profane talk, and give alms to the destitute; who restrain their carnal desires except with their wives and slave-girls [literally: "what their right hands possess"], for these are lawful to them... (*Quran* 23:1–6)

The *Quran* not only condones people mastering other people as their property, it also forbids masters to deny the bounty the Lord has bestowed on them by making their slaves equal partners in their wealth:

> To some God has given more than He has to others. Those who are so favored will not allow their slaves an equal share in what they have. Would they deny God's goodness? ...On the one hand there is a helpless slave, the property of his master. On the other, a man on whom We have bestowed Our bounty, so that he gives of it both in private and in public. Are the two equal? God forbid! (*Quran* 16:71, 75)

Though there are passages in the *Quran* that commend the freeing of a slave as a charitable deed, slavery is not outlawed as an institution. Such can be the only conclusion drawn from passages that mention slavery in order to set conditions for its proper practice.

One might be consoled by the thought that even if Islam is irremediably misogynistic and oppressively draconian, a Muslim might at least opt to leave her or his religion — a right which seems to be protected by Article 18 of the UDHR. This brings us to Saudi Arabia's second cited ground for abstention from ratification of the UDHR. The delegation of Saudi Arabia argued that the "freedom to change his religion or belief" granted in one of the clauses of Article 18 should be omitted because it exemplified the kind of thinking that led to the crusades and religious wars.

While this Orwellian pseudo-argument was supported by delegations from fellow muslim countries such as Iraq and Syria, its absurdity was pointed out by the Philippines' representative, who correctly reversed the argument to the effect that: "the precedents of the crusades and the wars of religion... clearly demonstrated the utility of provisions designed to prevent repetition of such conflicts."[71] That the Saudi delegation's objection to this point was on Islamic grounds is emphasized by Baroody asking the French representative

> whether his Government had consulted the Moslem peoples of North Africa and other French territories before accepting that text, or whether it intended to impose it on them arbitrarily. He also asked the other colonial powers, notably the U.K., Belgium and the Netherlands, whether they were not afraid of offending the religious beliefs of their Moslem subjects by imposing that article on them. He reminded the representative of Lebanon that 40 percent of the Lebanese population was Moslem, and asked him whether the whole of that population had authorized him to approve Article 18 *in toto*.[72]

Representatives from Muslim nations were not the only delegates to recognize that freedom to disown one's religion was inadmissible to Muslims. Shockingly, the delegate of Denmark, Bodil Begtrup, abstained from voting on Article 18 because "the adoption of the second phrase would mean that the representatives of 300 million Mohammedans [now 1.3 billion] would be unable to support the declaration."[73] Indeed, the *Quran* mandates that voluntary conversion out of Islam, or 'apostasy', is punishable by death unless the apostate repents:

> *Prophet, make war on the unbelievers and the hypocrites* and deal rigorously [*jihad*] with them. Hell shall be their home: an evil fate. They swear by God that they said nothing. Yet *they uttered the word of unbelief and renounced Islam after embracing it*...If they repent, it will indeed be better for them; but if they give no

---

71 Morsink, *The Universal Declaration of Human Rights*, 25–26.

72 Ibid., 26.

73 Ibid., 25.

heed, God will sternly punish them, both in this world and in the world to come. *They shall have none on this earth to protect or help them.* (9:73–9:75)

It is in this context that we should view the *Quran*'s verse 2:256 claiming that: "There shall be no compulsion in religion." It does not mean that *once having accepted Islam*, one is then free to renounce it. One is not allowed to change one's mind because one has learned more than one knew when first having called oneself a "muslim". Disowning Islam is considered a kind of betrayal of the community of Allah that is clearly punishable by death. Furthermore, while conversion to Islam is not mandatory, the only unbelievers that are tolerated are members of other Abrahamic religions (Judaism and Christianity) who have been utterly subjugated by Muslims, and have accepted Islamic authority as legitimate by paying Muslim authorities a hefty non-believers' tax:

> Fight against such of those to whom the Scriptures were given as believe neither in God nor the Last Day, who do not forbid what God and his apostle have forbidden, and do not embrace the true Faith, until they pay tribute out of hand and are utterly subdued. (9:29)

The *Quran* takes great pains to make clear that its injunctions are perfect, eternally valid, and are to be followed without any alteration. It completely denies the possibility of reformation in the context of a democratic society. Islam's incompatibility with a legal order based on changing popular will is made clear in verses 6:114–116, which depict the *Quran* as a perfect and complete guide to life, including perfect legal "justice", *that should be followed over the opinions of the majority of people in the world*:

> Should I seek a judge other than God when it is he who has revealed the Book for you with all its precepts? Those to whom we gave the scriptures know that it is the truth revealed by your Lord. Therefore have no doubts. Perfected are the words of your Lord in truth *and justice*. None can change his words. If you obeyed the greater part of those on earth, they would lead you away from God's path.

Verses 43:2 and 85:21–22 both clearly state that the *Quran* is a literal transcript of an "eternal book" inscribed on an "imperishable tablet" in

God's keeping: "We have revealed the Koran in the Arabic tongue that you may understand its meaning. It is a transcript of the eternal book in Our keeping, sublime, full of wisdom" (43:2); "Surely this is a glorious Koran, inscribed on an imperishable tablet (85:21–22)."

If one has any remaining doubt as to the intended "eternal" validity of the verses of this book inscribed on an "imperishable" heavenly tablet and merely translated for Muhammad, verses 86:12–14 aim to leave us with no doubt whatsoever: "By the sky that thunders, by the earth that splits, this [Qur'an] is a word once and for all, not meant lightly." Verse 2:85 insists that the *Quran* must be followed in the entirety of its injunctions, which means that whether or not one has a more profound mystical inner faith one is still bound by its legal precepts: "Can you believe in one part of the Scriptures and deny another? Those of you that act thus shall be rewarded with disgrace in this world and with grievous punishment on the Day of Resurrection."

It is echoed by verses 2:174–177: "Those that suppress any part of the Scriptures which God has revealed in order to gain some paltry end shall swallow nothing but fire into their bellies…That is because God has revealed the Book with the truth; those that disagree about it are in extreme schism." Finally, the very first line and opening verse of the *Quran*, prefacing its contents, explicitly declares: "This book is not to be doubted." (2:1) As Verse 6:36 attests, the legal authority of this indubitable book is comprehensive: "We have left out nothing in the Book." Muhammad's last revelation in verse 5:3: "This day I have perfected your religion for you and completed My favour to you," means that Islam, as defined by the content of the *Quran*, was perfected at that time in such a way that any historical evolution is precluded. The *Quran* also disallows the possibility that falsehoods could enter into its text over time (such that its authority might be questioned on the grounds of 'historical corruption'): "This is a mighty scripture. Falsehood cannot reach (enter) it from before (the past) or from behind (the future)." (41:42) This is reiterated by the claim that: "You shall find no change in the ways of God." (33:62) The *Quran*'s claims concerning its eternally infallible status apply to the social order that it mandates, as well as the legal code of crime and punishment that it elaborates: "God lays down for mankind their rules of conduct." (47:3)

The Imperial Iranian government of Shahanshah Aryamehr Mohammad Reza Pahlavi was criticized by human rights groups for carrying out torture of political prisoners and inhibiting freedom of the press, freedom of assembly, and so forth.[74] The Carter administration eventually caved in to pressure from human rights activists and warned the Shah's government that these practices could not continue. When the Shah declared martial law during the initial riots and general strike of 1978, the Carter Administration refused to supply the Shah with rubber bullets, tear gas, and other instruments for maintaining order. Rather, they put pressure on Pahlavi to undertake 'democratic reforms', and his ultimate agreement to do so weakened the government in such a way as to result in the triumph of the popular Islamic Revolution. Under the Shah's so-called 'dictatorship' women and religious minorities enjoyed far more rights and liberties than after the populist Islamic Revolution of 1979.[75] In Pahlavi Iran, women served as members of the judiciary, even as Supreme Court judges. Religious minorities not only enjoyed full freedom of religion under the Shah, but were appointed to some of the highest level cabinet positions.

Shortly after the popular Revolution of 1979, a revolution against 'dictatorship' encouraged by Western human rights activists in the name of 'democratization', the Islamic Republic of Iran instructed its new ambassador to the United Nations to disown Iran's commitment to the *Universal Declaration of Human Rights*. The following is a paraphrase of the comments of Iranian Ambassador Said Raja'i-Khorasani from the official record of the UN General Assembly, in his speech to that body's 39th session on Friday December 7th, 1984 at 3pm in New York:

---

74 Nikki R. Keddie, *Roots of Revolution: An Interpretive History of Modern Iran* (New Haven: Yale University Press, 1981), 233–234.

75 Reza Pahlavi, *Winds of Change: The Future of Democracy in Iran* (New York: Regnery Publishing, 2002).

> The new political order [of the Islamic Republic] was... in full accordance and harmony with the deepest moral and religious convictions of the people and therefore most representative of the traditional, cultural, moral and religious beliefs of Iranian society. It recognized no authority... apart from Islamic law... conventions, declarations and resolutions or decisions of international organizations, which were contrary to Islam, had no validity in the Islamic Republic of Iran... The Universal Declaration of Human Rights... could not be implemented by Muslims and did not accord with the system of values recognized by the Islamic Republic of Iran; his country would therefore not hesitate to violate its provisions.[76]

Evidently the representative of the Islamic Republic of Iran forgot what his country agreed to when it joined the United Nations, as a founding member. The UN Charter which was signed on 26 June 1945, in San Francisco, begins with these words:

> We the peoples of the United Nations determined to save succeeding generations from the scourge of war, which twice in our lifetime has brought untold sorrow to mankind and to reaffirm faith in fundamental human rights, in the dignity and worth of the human person, in the equal rights of men and women... do hereby establish an international organization to be known as the United Nations.

The world had seen many small regional conflicts in the lifetime of the drafters of this Charter, so that the word "twice" employed in this Preamble can only be a reference to the first and second *world* wars, which were indeed the cause of unprecedented human misery and destruction on a global scale. The many smaller conflicts that have occurred since these words were first proclaimed in San Francisco that summer have not yet amounted to a failure of the organization to achieve its primary objective. That objective has always been, and remains, the prevention of a Third World War — a war that is not merely regional, one whose consequences are global, one wherein weapons of mass destruction may again be employed against civilian populations as they were in Hiroshima and Nagasaki at the close of the last world war. These opening words of the

---

76   Elizabeth Ann Mayer, *Islam and Human Rights* (Boulder: Westview Press, 2007), 9.

Preamble also state unequivocally that the reaffirmation of "fundamental human rights" including "the equal rights of men and women" is integral to the organization's mandate to prevent another world war.

This is reinforced by Article 1:3, which defines "promoting and encouraging respect for human rights and for fundamental freedoms for all without distinction" as one of the "purposes of the United Nations". While Article 2 of the Charter grants that the UN "is based on the principle of the sovereign equality of all its Members", it also demands that "All Members, in order to ensure to all of them the rights and benefits resulting from membership, shall fulfill in good faith the obligations assumed by them in accordance with the present Charter." As noted, "promoting and encouraging respect for human rights and for fundamental freedoms" is one of these obligations assumed by any nation-state upon membership. Finally, the sixth section of Article 2 boldly proclaims that: "The Organization shall ensure that states which are not Members of the United Nations act in accordance with these Principles so far as may be necessary for the maintenance of international peace and security."

It should be no surprise that, ironically, by means of the doctrine of mutually assured destruction (M.A.D.) the United States and the Soviet Union prevented another world war from happening for almost half a century after the adoption of the UN Charter. It is only now, in a world destabilized by the consequences of the disintegration of the Soviet Empire, that we again face a Third World War — with Islam. Khorasani's legitimately Islamic withdrawal of Iran's pledge, as stipulated by recital 6 of the UDHR Preamble, to "achieve, in co-operation with the United Nations, the promotion of universal respect for and observance of human rights and fundamental freedoms" shatters the one idea fundamentally underlying the Freedom of Religion protected by Article 18 of the *Universal Declaration of Human Rights*. This discredited idea is that *any and all religions can peacefully coexist within a single state without compromising the security of that state*, because *the political power defining the state can be "neutral" with respect to religion.*[77] In his landmark 50th anniversary

---

[77] Morsink, *The Universal Declaration of Human Rights*, 259–260.

study of the "Origins, Drafting and Intent" of *The Universal Declaration of Human Rights*, Johannes Morsink contends that though the drafters of the UDHR recognized that it embodied a certain ideology, they believed that this was a "neutral ideology."[78]

According to this ideology, the foundation of our inalienable human rights lies in our each being endowed with reason and conscience.[79] The drafters believed that the UDHR embodied absolute moral rights that preexist any legislative or judicial acts of governments or social conventions and traditions — all of which are, by comparison, merely accidental circumstances of history.[80] The UDHR was prompted by the Second World War, a historical circumstance indeed, but its drafters used the words "reaffirm" and "disregard and contempt" in respect to human rights to make clear that the principles of the Declaration were atemporally rooted in attributes of a metaphysical human nature, and that these principles were wantonly violated during World War II. The Preamble phrase "whereas a common understanding of these rights and freedoms" was proposed by representative Malik of Lebanon, as an indication of the rudimentary but fundamental metaphysical consensus of the drafters.[81] Morsink summarizes this metaphysical consensus of the drafters of the UDHR in the following terms in the course of refuting a utilitarian interpretation of the Declaration:

> ...it is going too far to say that the votes on some philosophically laden terms do not represent any philosophical consensus. *Metaphysically, the great majority of the drafters accepted the view that human rights are inherent in people and therefore inalienable moral birth rights.* And epistemologically the great majority of them accepted conscience and reason as ways of knowing that lead us into that realm of inherent moral rights. These beliefs form a minimal but widespread consensus...

---

78 Ibid., 327.

79 Ibid., 284–302.

80 Ibid., 295.

81 Ibid., 317.

I conclude that while *the drafters* surely thought that proclaiming this Declaration would serve the cause of world peace, they *did not think of the human rights they proclaimed as only or merely a means to that end*. Regardless of the consequences for world peace, *these rights have an independent grounding in the members of the human family* to whom they belong and who possess them as birthrights.[82]

The metaphysical foundation of the UDHR is most evident in the recitals of its Preamble and in Articles 1, 2, and 7. According to Article 1 of the *Universal Declaration of Human Rights*: "All human beings *are born free and equal* in dignity and rights. They are *endowed* with reason and conscience and should act towards one another in a spirit of brotherhood." That all human beings *always already* have "equal" rights despite any factual or material differences between sexes, races, or individuals of biologically unequal intellectual or physical capabilities (including the retarded and deformed), indicates that a *metaphysical human being* is the subject of the rights elaborated in the UDHR. That human beings are said to be "born free" in addition to being born equal, is a reference to the fundamental free will of each individual, which alone allows each person to be a morally responsible agent "endowed" with genuine "reason and conscience".

Compare this to the fatalism of Islam, which radically denies the free exercise of individual conscience and does so in such an irrational manner as to also preclude any rational objections to Allah's revealed will. Note these verses of the *Quran* that explicitly assert predestination of the human soul, and the lack of morally responsible free agency:

> If God wills to guide a man, He opens his bosom to Islam. But if he pleases to confound him, He makes his bosom small and narrow as though he were climbing up to the sky. Thus shall God lay the scourge on the unbelievers.

> …Their idols have induced many pagans to kill their children, seeking to ruin them and to confuse them in their faith. Had God pleased they would not have done so.

(6:125,137)

---

82 Ibid., 318; my emphasis.

We have cast veils over their hearts and made them hard of hearing lest they understand your words. They will believe in none of Our signs, even if they see them one and all.

…Had God pleased He would have given them guidance, one and all.

…Such is God's guidance; He bestows it on whom he pleases of His servants.

(6:25, 35, 88)

God alone points to the right path. Some turn aside, but had He pleased, He would have given you guidance all.

…Strive as you may to guide them, God will not guide those whom He confounds. There shall be none to help them.

…Had God pleased, He would have united you into one community. But He confounds whom He will and gives guidance to whom He pleases. You shall be questioned about all your actions.

…God does not guide the unbelievers. Such are those whose hearts and ears and eyes are sealed by God; such are the heedless. In the life to come they will surely be the losers.

(16:9, 37, 93, 107–109)

Indeed, the wrongdoers are led unwittingly by their own appetites. And who can guide those whom God has led astray? There shall be none to help them.

…He does not love the unbelievers.

(30:29, 45)

God leaves in error whom He will and guides whom He pleases.

(35:8)

He whom God confounds shall have none to guide him.

(40:33)

As for the unbelievers, it is the same whether or not you forewarn them; they will not have faith. God has set a seal upon their hearts and ears; their sight is dimmed and grievous punishment awaits them.

(2:6–7)

> Think! Who, besides God, can guide the man who makes his lust his god, the man whom God deliberately confounds, setting a seal upon his ears and heart and drawing a veil over his eyes?
>
> (45:23)

According to these passages from the *Quran*, the mysterious and unfathomable "grace" of the Revealer must be ultimately responsible for both faith and disbelief. If individuals were in the last analysis responsible for acquiring faith *themselves*, i.e. *apart from* God, this would mean that Revelation is not above the need for rational justification. Yet, even if the Revealer is really the ultimate source of unbelief, impious individuals who defy the will expressed in his Revelation can nonetheless be legitimately judged by him. Allah predestines certain people to do evil and then damns them to suffer eternal torture in hell on account of having committed such evil acts. To call this incoherent or self-contradictory is to assert the priority of the rationality of individuals over the will of an omniscient and omnipotent God who, as such, is unfathomably mysterious. Merely human reason and conscience are also denigrated in the *Quran*'s rejection of men's conscientious objection against taking up arms:

> Fighting is obligatory for you, much as you dislike it. But you may hate a thing although it is good for you, and love a thing although it is bad for you. *God knows, but you know not.* (2:216)

The debt of the UDHR to Enlightenment political philosophy is clear, and was even acknowledged by Representative Chang of the Republic of China who urged the drafting Committee to "build on the work of the eighteenth century philosophers."[83] With the words "determined to promote social progress and better standards of life in larger freedom", Recital 5 of the Preamble implicitly recognizes previous attempts by Enlightenment thinkers such as the Marquis de Condorcet to have the "equal rights of men and women" recognized, and it promises to succeed where they failed.

---

83 Ibid., 287.

One would imagine that the drafters of the UDHR would have realized that this endeavor might *fundamentally* conflict with certain 'revealed' religious traditions, especially given the metaphysical or spiritual nature of the Declaration's claims regarding "reason and conscience". Alas, for whatever reason, enough of them failed to realize this so as to allow for an unqualified formulation of Article 18's right to freedom of religion. In his study of the UDHR drafting debates, Morsink summarizes the "neutral" outlook which the drafters believed that they were taking with specific reference to the listing of religion among the nondiscrimination items in Article 2, when he writes:

> The Declaration rejects every connection between religion and politics... There is no presumption in the Declaration that the morality of human rights requires any kind of religious foundation... Most groups and nations that see an intimate connection between religion and the machinery of the state do so because they think that religion is needed to provide the moral cement that holds a nation together. The Universal Declaration rejects this connection and gives everyone total freedom of religion, including the right not to have one. It places the religious convictions and expressions of all persons on the same equally distant footage from the seats of political power.[84]

Habib Malik, the Lebanese delegate, summed up the spirit of Article 18's extraordinary claim when he said that it was based on the Human Rights Committee's recognition of "the fundamental right of *differing fundamental human convictions, as in religion*, to exist in the same national entity", meaning that "a single nation is obligated by international law, to recognize *the diversity of fundamental points of view on ultimate matters*."[85] In other words, Article 18 gives free reign to the practice of a religion involving non-negotiable, absolute, all-grounding, ultimate criteria for all judgments in life.

Any and *all* possible 'religions' involving "fundamental points of view on ultimate matters" would have to be acceptable, because if one begins to say that this or that religion does not qualify for toleration, then the

---

84 Ibid., 263.

85 Ibid., 260; my emphasis.

political power of the state *is not neutral* with respect to religion. This means that if it can be demonstrated that any one religion with a considerable following explicitly conflicts with the ideology of the *Universal Declaration of Human Rights*, then the claim of the UDHR that it is possible to establish a form of political power that is "neutral" with respect to religion is false. I have made just such a demonstration with regard to Islam, the fastest growing religion on Earth — projected to be the religion of the majority of our planet's population within this century.

CHAPTER 2

Planetary Emergency

Perhaps if Universal Human Rights cannot serve as the basis for a global geopolitical order, then the universalization of democracy might be able to do so? But what does "democracy" even mean, and how is it related to the ideology of "liberalism"? If the core values of certain civilizations, such as Islam, are inherently illiberal, then how can "liberal democracy" be universalized while remaining democratic? In *The Crisis of Parliamentary Democracy* and other writings, Carl Schmitt uncovers the conservative origins of Democracy and suggests that this political form is ultimately incompatible with Liberalism. A look back at the cultural and historical context of Democracy in classical Athens as well as its purest modern reconstruction by Jean-Jacques Rousseau supports Schmitt's claim that the kind of liberal political ideology epitomized by John Stuart Mill's *On Liberty* is indeed radically incompatible with Democracy. Consequently, the contemporary prevalence of so-called 'Liberal Democracy' is based on a profound confusion of political concepts and a conflation of potentially opposed interests that took shape in a very particular historical context. Both Democracy and Liberalism are problematic in themselves, and taken together they constitute a monstrous hybrid that cannot be expected to endure any serious stresses.

Democracy has its roots in the so-called 'Athenian Revolution' of the 460s and 450s BCE, which stood for the "rule of law" in opposition to a tyrant's unbridled power to arbitrarily do what he wishes. However, reformers like Perikles never meant to change the theological basis of law or institute an unqualified "majority rule".¹ In fact, political participation and civil rights in Greek democracy never extended beyond a given male descent-group to include women, slaves, or foreigners (resident aliens).² These latter three groups were considered 'outside' the community even though, when taken together, they formed the overwhelming *majority* of a city-state's population.³ Indeed, the Greek word for tribe or clan is *deme*, and this is the basis of the term *democracy*. In other words, "democracy" does not refer to "the people" in the sense of a mass of human beings in general. Rather, it is a specific reference to one's male tribal kinship affiliations. The basic sense of the word is therefore that no oligarchs or sovereigns should set one against one's own kinsmen.

Legally, women in Athenian democracy were considered property of one kind or another. The established view was that women were by nature made to serve men in various ways, some being fit for producing legitimate heirs, others for housework, yet others for sexwork—with a clear and inviolable division between the three.⁴ The role of the housewife was to be properly trained by her husband to look after his domestic property and supervise his household servants while he attended to public business or private pleasure outside the home.⁵ Whereas it was both natural and legal for married men to indulge in sexual relations with prostitutes, the chastity of citizens' wives was strictly guarded in order to ensure that any male children could be legitimate heirs of their husbands' estate or that any female children could be legitimately sold off as brides to other male citizens.⁶ If it ever came about that the wife of a citizen was the sole heir of his property (because he had no male children or they had died), she was forced to marry his nearest male relative so as to immediately transfer control of the property in question back to (a male member of)

---

1   J.K. Davies, *Democracy and Classical Greece* (Cambridge: Harvard University Press, 1993), 114.
2   Ibid., 25.
3   Ibid., 35.
4   Ibid., 219.
5   Ibid.
6   Ibid., 219–220.

her husband's family. Part of this chastity was modest dress, and a general expectation that they would stay within the 'covered' domain of the household.[7]

Female domestic slaves were even worse off than the wives of citizens and could suffer domestic abuse at the hands of these frustrated housewives. The only women in this democratic society that had any de-facto (not legal) autonomy at all were prostitutes or *hetairai*, who earned a salary of their own.[8] Rich and powerful men enjoyed a closer companionship with these more sophisticated 'escorts' than with their own wives, although it was taken for granted that true friendship could only be between male citizens. The down side of this was that *hetairai* usually began as child prostitutes and only after years of training by a madam would they be able to see the wider world through providing companionship and sexual favors to upper class men.[9] Still, the lower the social class a woman belonged to, the more free she was to go out in public.[10]

The Athenian economy was fundamentally based on the institution of slavery. Slaves mined the silver from which coins were forged to pay for the corn imports without which Athens would starve.[11] Agriculture on any estate of a decent size was accomplished by slave labor, and slaves also manufactured domestic utensils, weapons, and other basic tools of civilized life.[12] They helped to build temples and engaged in morally suspect but necessary trade and retail activities. It is often believed that Greek 'slaves' were treated more like domestic servants that were part of the family. However, easy familiarity with their masters actually went hand in hand with casual subjection to violence, from which they had no legal protection or recourse.

---

7   Ibid., 219.
8   Ibid., 219–220.
9   Ibid., 217–219.
10 Ibid., 221.
11 Ibid., 89.
12 Ibid., 90.

Finally, Greek democracy lacked a distinction between religion and the state. This can be seen most clearly in the conservative motives behind, and limits to, educational reform in 4th century Athens. By the fourth century BCE, two developments threatened traditional Greek society. First, an education based on the poetry of Homer and the maxims of Pindar, on athletics, martial training, music and dance, had simply become inadequate for managing an increasingly complex society and state.[13] Second, geographer-historians like Herodotus had made Greeks aware of the very different traditional customs of foreign people, including the highly civilized Egyptians and Persians.[14] Reformist educators such as Isokrates opposed both traditional poetic and martial education and the view that all civil law and public morality were mere conventions and necessary pretenses that went against nature. Instead he argued that the faculty of reason, the inner *Logos* which guided abstract thought and argument, distinguishes men from animals and that it should be the basis of civil society.[15] Isokrates was considered controversial and was routinely harassed by traditionalists and sophists for these innovative and abstract notions.

However, *even Isokrates did not dare challenge the fundamental religious convictions of Greek society*. Rather, he defended them zealously against what he perceived to be the moral degeneracy of his time. He believed that it was incumbent upon public officials "that they should neither abolish any of the ancestral observances nor add anything which was not traditional" and he argued that piety consisted not "in vast expenses but in changing nothing of what they had inherited from their forefathers."[16] Once intellectual and oratorical education for pay, of the kind advocated by Isokrates, had become the norm, it benefited those who could afford to pay and the long-term effect of this educational reform was to reground conservatism.[17]

---

13 Ibid., 160.

14 Ibid., 107.

15 Ibid., 160–161.

16 Ibid., 171.

17 Ibid., 178.

Rituals and sacrifices were still incumbent upon all, regardless of whether the arcane mythologies in which they were rooted still made any sense.[18] Atheism existed in intellectual circles and was tolerated, but only in so far as these views were limited to rhetorical cynicism and did not threaten to change religiously based traditional social institutions in any way.[19] It was because Socrates was perceived to have crossed this line that he was condemned to death by the vote of a democracy, on charges of impiety and corrupting the morals of the youth.[20] There cannot be any more blatant testimony to the conservative theocratic character of democracy at its origin, than the impiety trial and public execution of Socrates in 399 BCE.

The severely antidemocratic sentiment of the *Republic* must have been much inspired by Plato's bitterness over the execution of his mentor. Beginning with Plato, and continuing through Cicero — who deeply influenced Jefferson and other founding fathers of the United States — an anti-democratic philosophical tradition of *Natural Right* arose that eventually culminated in the idea of "the inalienable rights of man". Most of the founding fathers of the United States were fiercely opposed to democracy, and they took care to build anti-democratic safeguards into the legal structure of their Constitutional Republic. Thomas Jefferson, author of the *Declaration of Independence*, with its "inalienable rights" grounding the constitutional order of the United States, took this view: "A democracy is nothing more than mob rule, where fifty-one percent of the people may take away the rights of the other forty-nine." Benjamin Franklin wrote: "Democracy is two wolves and a lamb voting on what to have for lunch. Liberty is a well-armed lamb contesting the vote." John Adams believed that: "Democracy… wastes, exhausts, and murders itself. There is never a democracy that did not commit suicide." James Madison, who argued decisively against democracy and in favor of a federal *republic* in the Federalist Papers # 10, had this to say: "Democracy is the most vile form of government… democracies have ever been spectacles of turbulence and

---

18   Ibid., 171.

19   Ibid., 172.

20   Ibid., 173.

contention... incompatible with personal security or the rights of property." The United States is commonly taken to be a "liberal democracy". However, it was intended to be neither "liberal", nor "democratic". The United States is, rather, by law a Constitutional Republic grounded in the protection of the "inalienable rights of man" as granted by a Creator God. Liberal democracy is a contradiction in terms.

The German political philosopher and influential Weimar jurist, Carl Schmitt, astutely recognized that the political philosophy of *liberalism* is not only separable from democracy, but may enter into conflict with it. In the view of Carl Schmitt the two systems of liberalism and democracy were forged together by a century of common struggle against monarchy. Democrats did not so much believe in open discussion on its own merit, but rather as a means to expose and oppose the secret machinations of princes. Just as liberals, otherwise weary of the intolerant uneducated rabble, adopted the slogan of egalitarian populism in order to co-opt masses united by an opposition to "royal absolutism". Once the common enemy is defeated, liberals and democrats come into conflict with one another and, according to Schmitt "as soon as it achieves power, liberal democracy must decide between its elements."[21] Schmitt epitomizes the crux of the conflict when, in *The Crisis of Parliamentary Democracy*, he writes: "The equality of all persons as persons is not democracy but a certain kind of liberalism, not a state form but an individualistic-humanitarian ethic and *Weltanschauung*. Modern mass democracy rests on the confused combination of both."[22] Upholding the rights of all, including the intolerant, to freedom from cruel and unusual punishment, freedom of belief, of speech, and the press, requires denying intolerant persons the right to pass binding legislation — even if they are a numerical majority.

Schmitt's critique of liberalism grew out of his experience of German politics in the Weimar period. According to the liberal democratic ideology of the Weimar Republic, even the most radical and anti-republican parties were allowed equal rights to run for election to the parliament.

---

21 Carl Schmitt, *The Crisis of Parliamentary Democracy* (London: MIT Press, 1988), 15.
22 Ibid., 12.

Upon winning a majority, they could act to abolish the constitution, denying all others the same equal rights to representation and participation that allowed them to come to power in the first place. Schmitt saw the Weimar experience as a paramount example of the internal contradiction of "liberal democracy", one wherein democracy or rule of the majority, may contradict the principles of liberalism enshrined within a constitution.

Carl Schmitt cites the political philosophy of John Stuart Mill as the paragon of Liberalism, and so it is to Mill's seminal text *On Liberty* that we must turn, bearing in mind all that has been said about the profoundly conservative and culturally conditioned character of Democracy at its origin in classical Greece. Mill's text *On Liberty* begins with an acknowledgment that when popular government was exceedingly rare and while it was still struggling to overcome nearly universal despotism or tyranny, it was hard to imagine that there would need to be any check on "rule by the people". This changed with the establishment of the United States of America. In this context it became clear that rule by the people does not mean the rule of each person over himself, but of a part of the people over another part, and over the individual.

Measures were taken by the founders of the United States to guard against "tyranny of the majority" but, according to Mill, safeguards against purely political tyranny over the individual are insufficient. The more insidious influence of mass *society* over "the soul" of the individual must also be checked, if liberty is to have any concrete significance.[23] Historically, those who have defied the opinions of society — and the laws based on those opinions — have only sought to secure general acceptance of their own deviations. They have not challenged, on principle, the right of society to legislate over the 'heretics' in the minority. Those who have dissented from one moral doctrine and are persecuted for it, once their different moral doctrine prevails, often become persecutors in its name.[24] Liberalism is, by contrast, an alliance of heretics against the mob — at least this is what Mill would have us believe.

---

23 John Stuart Mill, *On Liberty* (New York: Bobbs-Merrill, 1956), 6–7.

24 Ibid., 29.

Mill advocates "absolute freedom of opinion and sentiment on all subjects, practical or speculative, scientific, moral, or theological" irrespective of the opinions of others who believe our views, or our practical conduct in accordance with our opinions, to be "foolish, perverse, or wrong". He also extends this to the "freedom to unite [with others of the same views] for any purpose not involving harm to others [with different views]", given that those who combine their liberty are of full age and do so consentingly.[25] Mill argues that society only has a right to intervene in the life conduct of individuals, whether by force or by coercion, in order to prevent one individual from causing harm to another. To coerce or force someone to do anything because one believes it is for that person's "own good", physically or morally, is not justified: "Each is the proper guardian of his own health, whether bodily *or* mental and spiritual."[26] One may only try to persuade or reason with someone to 'better themselves' in such and such a way, but it is up to them whether they deign to listen to or completely disregard one's advice or admonishment. Advice, instruction, persuasion, and avoidance are the only means which others in society may use to disapprove of any conduct of an individual that concerns himself and does not harm others.[27] According to Mill: "Over himself, over his own body and mind, the individual is sovereign."[28]

Mill furnishes four reasons for freedom of thought and expression, or the free debate of ideas. Firstly, unless one claims infallibility, a suppressed view may be the truth on some matter; if it is not allowed to be examined, we cannot know with any confidence that this is not so. Secondly, it is very rare that a 'truth' is the whole truth on any given subject matter, and so free exchange and critical examination of ideas in the public forum is also necessary in order to allow something that is partly true to be complemented and rounded out by some other partial truth — and vice versa. Thirdly, even if a truth is known, at some point and by certain individuals,

---

25 Ibid., 16.
26 Ibid., 17.
27 Ibid., 114.
28 Ibid., 13.

to be true and to be the whole truth, it needs to be challenged (even if only by an artificial counter-argument) if it is not to be believed by most people on account of mere prejudice and without understanding its grounds. Fourth, 'truths' long accepted to be true and not any longer subject to question eventually atrophy and people lose the kind of conviction that one has when one arrives at a truth through personal experience; this by itself can cause such truths to be compromised in favor of ideas with more vitality (even if they are not as 'true').[29] Furthermore, Mill believes that persons should be able to act in accordance with their thoughts and words, so long as they do not become a "nuisance" to others and without "molestation" of others.[30] The same kinds of reasons that warrant freedom of thought and expression, also make it "useful" for people to be exposed to "different experiments of living" which exemplify "varieties of character" so that the worth of each of these ways of life should be tested and proved practically.

Mill goes on to make a series of what seem to be new claims distinct from those used to argue for freedom of thought and expression (in speech, the press, etc.). He believes that almost everyone lives in perpetual self-censorship, only thinking, speaking, acting, *and even desiring* as the masses desire, and never doing "what is not done" by *them*.[31] There is a tacitly fixed range of acceptable likes and dislikes (chess, horseback-riding, etc.). Exceptions for 'eccentric' behavior are made only for wealthy men, and even for them only up to a point, since they must fear their property being avariciously confiscated from them with the excuse of their being declared of unsound mind.[32] According to Mill, the human mental and moral faculties are as much in danger of being stunted or atrophying if they are not developed as the muscular powers of the body are if they are not exercised. The faculties of perception, judgment, discriminative feeling, moral preference, and firmness of self-control to carry out deliberate

---

29 Ibid., 64.

30 Ibid., 68.

31 Ibid., 74.

32 Ibid., 83.

decisions can only be developed and maintained if persons are regularly required to make their own choices. Unless she is forced to decide for herself between different possible courses of action, and thereby to devise her own "plan of life", a person could do without any faculties other than those required for "ape-like... imitation".[33] Mill claims that human beings *are*, as a matter of fact, each different from one another in the constitution of their faculties and inclinations. They are not like sheep, and even sheep are not all alike. He argues that if each individual is allowed to develop himself or herself, each will be more self-motivated and enthusiastic, and this will bring not only greater diversity but greater vital energy to the collective of humanity — strengthening the collective by making it more dynamic.[34]

The right to live originally, without being persecuted by mass society, is a precondition of geniuses arising in a given culture. Mill believes that although most people superficially appreciate the talent it requires to paint a fine painting or write a fine poem, true originality in thought and action is at the very least implicitly viewed by them as something that one could do without. If people could see what originality could do for them, either it would not be originality — or they would themselves be producing works of genius (which is to say the same thing, since, by definition, not everyone can produce works of genius). Mill insists emphatically on the importance of genius.[35] This is not argued for specifically in *On Liberty*, so we may assume that it is an extension of his claim that a multiplicity of life-experiments will contribute to the dynamism and thus the enduring vitality of humanity as a whole.

In respect to genius, the example of Socrates, and the method of questioning named in his honor, plays a central role in Mill's argument. Mill first mentions Socrates in connection to his insistence that no exception should be made for persecution of moral opinions — as opposed to those merely dealing with scientific matters of fact. The freedom to dissent from

---

33   Ibid., 71.

34   Ibid., 76.

35   Ibid., 79.

the accepted moral code of society is the most important of all. He reminds his readers that the charges brought against Socrates in the democratic assembly of Athens were his being guilty of impiety (disbelieving in the gods of the State) and corrupting the morals of the youth. This is one of the greatest instances of a generation acting rashly and regretting its error, once it becomes commonly acknowledged that the man executed was not only no mere criminal, but was even a paragon of ethical conduct.[36] However, for Mill, Socrates is not just another vindicated heretic.

Mill takes the Socratic method as exemplary of the kind of dialectical debate that every 'truth' should undergo before it is accepted as 'certain'. He mentions this while lamenting the distant day when all discoverable truths have been widely accepted as certain, suggesting that at such time it will be vital that the various positions opposed to these 'truths' be retained and revived with as much sincerity as possible by dialogue partners. He takes the Socratic dialogues of Plato to have had this aim, namely to set all the great philosophical positions against one another, so that no one of them is affirmed without its full significance being brought out through critical debate with the others. He notes that in his time, there are many critics of purely negative dialectics that use logic to undercut various positions without replacing them with any positive truth. Very significantly, Mill admits that if the result of such dialectical reasoning were always purely negative, it would be poor. He seems to think that negative dialectics always leads to "stable belief" that is at the same time not mere unquestioned commonplace opinion, and he takes this to be true of Plato's original Socratic dialogues as well.[37] Given the great effort it will take to reconstruct and convincingly argue for abandoned positions once most anything is certain, we should be grateful to still have genuine opponents to test our 'truths', because even if they prove to be correct, this will deepen our enthusiasm and our understanding of why we take them to be so.[38]

---

36 Ibid., 29–30.

37 Ibid., 54–55.

38 Ibid., 55.

However, there is a serious problem with this line of argument. It assumes that critical questioning and debate are always *determinatively* negative, in other words, that they will lead to the discovery of more solid *positive* truths, and moreover ones with ethical and political significance. Socrates was executed by the democratic assembly of Athens not primarily in order to prevent the arrival of new truths, but out of fear that his method of questioning was nihilistic and would ultimately (mis)lead the young and bold in his entourage to the conclusion that *nothing is true*.

Mill himself clearly rejects this negative conclusion as a possibility. This is apparent in his claim that although truth is fostered by testing itself against error, there will come a time when virtually all truths there are to be discovered will have been discovered and known to be certain.[39] That these certain truths are supposed by Mill to have a positive moral and political content is clear from a number of significant exceptions that Mill makes to the principle of harm being the sole criterion of legitimate State intervention in the lives of individuals. These exceptions largely concern offenses against "public decency". Note this somewhat veiled reference to (presumably) public nudity ("indecent exposure") or sexual conduct (especially masturbation), equating them with relieving oneself—an act which is necessary, and not condemnable, though one which he takes it to be obvious ought not to be performed before the eyes of others:

> Again, there are many acts which, being directly injurious only to the agents themselves, ought not to be legally interdicted, but which, if done publicly, are a violation of good manners and, coming thus within the category of offenses against others, may rightly be prohibited. Of this kind are offenses against decency; on which it is unnecessary to dwell, the rather as they are only connected indirectly with our subject, the objection to publicity being equally strong in the case of many actions not in themselves condemnable, nor supposed to be so.[40]

Furthermore, while Mill admits that "fornication" and gambling must be allowed, he claims that there are good arguments in favor of prohibiting public advertising of these establishments, as well as for penalizing pimps

---

39 Ibid., 53.
40 Ibid., 119.

and those who run gambling houses.⁴¹ He also believes in putting high taxes on alcohol consumption, since the state must tax something, and it is something that can best be gone without and something that, in excess, is a source of danger to the public.⁴² This series of statements attest to his holding as 'true' some very strong value judgments that are not argued for. One has to wonder whether these value judgments make Mill any different from the social majority that tyrannizes over persons with eccentric, but not strictly harmful, views.

For example, what rationale is it that allows public boxing matches, or other violent sports, but bans public sex between consenting individuals? What argument can be made against public nudity as "indecent exposure" that could not also essentially be used to argue for mandatory hejab in societies with different standards for how much of one's body being exposed is "decent"? Given his comments depreciating 'civilized' Europeans by comparison to much more dynamically alive and naturally "human" native cultures, Mill should not make an error like this. After all, the Nuba of Kau in Africa go about their towns and engage in all of their activities completely naked. Is advertising for a brothel any worse than handing out evangelical literature on a sidewalk, effectively promoting a Church that preaches a hateful and oppressive ideology? Why not tax sales of the Bible, since there is a good argument that mass consumption of the sadistic brutality on display in the Old Testament is far more damaging to public health than the consumption of alcohol?

If one were to object that these questions are too "extreme", then one would be violating Mill's own criteria for the protection of individual liberty, especially the liberty of the intellectual genius who questions everything. Mill argues against moderation when it is a question of freedoms that do not directly harm someone else. There is, according to him, no such thing as freedom to question received or generally accepted opinions and practices only *up to a degree*, and only insofar as what is questioned does not lie within the scope of that which is considered evident and most

---

41 Ibid., 120, 122.

42 Ibid., 123.

certainly true. So long as questioning society's norms, in thought, word, or deed, does not harm another person it is never "pushed to an extreme".[43]

Mill believes that great thinkers must uncompromisingly follow their lines of thought to whatever 'Truth' it may lead them, and that the errors produced thereby are worth more than truths affirmed on authority by those who have never thought through these propositions for themselves. A man of great intellect wastes his life and mind trying to sophistically render his thoughts compatible with orthodoxy.[44] Thus, in his admission of crimes against "public decency", Mill is violating his own criteria. If he is to consistently follow through with the premises of his liberal political theory, the definition of "harm" justifying state intervention narrows indefinitely. There will always be some thinker who can successfully argue that no direct harm is done to others by individuals who engage in public nudity, public sex, and even public defecation under certain circumstances (is someone who vomits in a public park arrested?). What about sex acts that involve violent treatment of others, or the holding of others in conditions of bondage or enslavement — albeit with their expressed consent (at least, initially)?

This brings us to the most significant exception that Mill makes to the liberty of the individual in so far as he does not harm others. This concerns an individual voluntarily entering into a contract whereby he permanently relinquishes his liberty. Mill claims that although when one enters into contracts it should generally be expected that one will abide by them, this cannot hold true of a contract whereby a person sells himself into slavery. Mill's argument here is that liberty is allowed to the individual so that he may be left to decide how best to better himself, but enslaving himself cannot be in any way considered doing good to himself. If a man relinquishes his freedom voluntarily, he relinquishes the rationale of his right to individual liberty at least in so far as he is free to sell himself into

---

43   Ibid., 26.

44   Ibid., 41.

slavery.[45] This he is *not* free to do, and no contract of this kind can be honored by law.

This passage is very significant, because it shows that the core of Mill's Liberalism is the idea that people should be forced to be 'free', even if they freely desire to be enslaved. What if a person freely believes that the experience of subjugation, of being bound and disciplined by others will purify his soul by mortifying his body and annihilating the ego that separates him from the divine? As Aldous Huxley reminds us in *The Devils of Loudun*, this would be very similar to the practice of some ascetic monastic orders of the middle ages. Mill violates the core of his own Liberalism by suggesting that a person would not be allowed to undertake such 'purificatory' masochism, which would entail — by previous contract — that if he cries out that he has had enough and begs to be released he positively should *not* be released, ever, until, he dies and — according to his own belief — enters the afterlife with his soul purer than it otherwise would have been.

Presumably, if a person is not allowed to sell herself into slavery, she is also not legally allowed to commit suicide. In other words, if others are aware of a person's intention to take her own life, they are allowed to legally intervene to prevent her from doing so — perhaps by putting her behind the padded walls of a mental institution or forcing her to be medicated out of the suicidal state of mind.

But imagine a person believes in reincarnation and thinks that suicide in *this* life at a particular time is justified with a view to being able to live at a particular place and time that would allow him to do better for himself and for others. In other words, suppose a man who is now 50 believes that he has to make a particular contribution to humanity, intellectual, artistic, or of whatever nature, in the year 2050. If he lives out his present life, he might just make it to 2050, but he will be too old to play that role. Or if he dies just before 2050, he will only be a boy in his subsequent incarnation, and this will also prevent him from playing such a role. To deny a man with these beliefs the freedom to end his own life when he so chooses, is

---

45 Ibid., 124–125.

to get government into the business of regulating valid beliefs — which is exactly what Mill claims to be fighting against.

However, if we admit that not only enslaving oneself, but even killing oneself is permissible, as a free choice, then there are no conditions on freedom. So long as one is not terribly incapacitated (and therefore also not the autonomous subject relevant to Mill's argument), then one can always choose to end one's life. Insofar as one does not make *this* choice, one is consenting to everything that one experiences in one's life — including all of the *harm* done to one by others. One has chosen to take the risk of living, and of learning from painful experiences, and of being strengthened by what harms us but does not kill us.

In a key passage from the introductory section of *On Liberty*, Mill equates Liberty with true freedom: "The only freedom which deserves the name is that of pursuing our own good in our own way, so long as we do not attempt to deprive others of theirs or impede their efforts to obtain it."[46] Very significantly the first sentence of *On Liberty* consists of a clarification that the essay is not about the "liberty of the will", or free will, which Mill apparently believes philosophers are wrong to oppose to the doctrine of necessity (determinism). In other words, Mill is a compatibilist who believes that political freedom — or "liberty" — is unaffected by the truth of causal determinism.[47] There is a connection between these two passages from the opening of *On Liberty*. There is no "so long as" where Freedom is concerned. Freedom is freedom; it is an existential reality that has no political conditions on it. Mill's Liberalism does not preserve this natural freedom.

By setting conditions on freedom, such as that one cannot sell oneself into slavery because it is to do to oneself what someone else considers a harm to oneself, Mill turns freedom into something granted to the individual by the government on the basis of a certain rationale. This rationale is determined by someone other than oneself, and it does not even have the majoritarian legitimacy of being the democratically expressed general

---

46 Ibid., 16–17.

47 Ibid., 3.

will of "the public". This follows from the fact that Mill believes there is *only* political freedom, since there is no freedom of the will that genuinely transcends natural determinism. In other words, no person is always already *free*, in so far as she has an autonomous will.

Jean-Jacques Rousseau believed that one cannot appeal to the conscious life of "natural man" for any rational principles of human conduct. Natural man is a blank slate and for this reason he is infinitely perfectible, but is also capable of infinite degradation. While Rousseau adopted this position in order to preserve an unconditioned ground for the experience of genuine existential freedom, it led him to an oppressively conservative political theory. If there is no *original or inherent human nature*, then all rights are social or civil rights granted by the sovereign authority of civil society and one cannot appeal to the state of nature or to natural rights in order to oppose civil laws of any kind. Civil society requires "the total alienation of each associate, with all his rights, to the whole community."[48] Having taken this to be the case, Rousseau believed that the way to preserve the greatest degree of natural freedom in civil society is to completely equate the will of each with the will of all. He developed the concept of a *general will* that is rational, i.e. universal or abstract, in the sense that it is what is left of the will of each person that would also be in the interest of every other person.

Rousseau recognized that this majority-rule would actually be universal disenfranchisement if each person had an entirely unique will that was not at all represented by the compromise of the majority vote. In other words the general will is not simply produced because any self-interest of each individual is canceled out by opposed self-interests of other individuals so that only the *common good* is expressed by the majority vote on some issue. The majority vote is not a compromise, it reflects a true *general will* that *is identical* with the individual will of right-minded and unselfish citizens. The minority vote represents a failure by a few citizens to unselfishly understand the common good of society in respect to a

---

48 Jean-Jacques Rousseau, *The Basic Political Writings* (Indianapolis: Hackett, 1987), 285.

certain issue. This minority cannot then go on to hold their wrong view, but must be enlightened as to the true *general will* by the outcome of the vote.

No man is fully a human being before civil society, and as a human being he is produced in his very "life and being" by the general will. To vote with what turns out to be the minority means to have misunderstood oneself. In the second chapter of *On the Social Contract*, which is entitled "On Voting", Rousseau writes:

> When a law is proposed in the people's assembly, what is asked of them is not precisely whether they approve or reject, but whether or not it conforms to the general will that is theirs. Each man, in giving his vote, states his opinion on this matter, and the declaration of the general will is drawn from the counting of votes. When, therefore, the opinion contrary to mine prevails, this proves merely that I was in error, and that what I took to be the general will was not so. If my private opinion had prevailed, I would have done something other than what I had wanted. In that case I would not have been free.[49]

Rousseau identifies two related sources for the ideological homogeneity required to produce the general will in the majority of the citizenry. Firstly, the tribe with its ancestral customs and unique cultural institutions grows up out of the state of nature in a spontaneous manner, conditioned by accidents of time and place that are different in the case of each tribe.[50] These ancestral customs and cultural institutions constitute a kind of 'national philosophy' or ethos.[51] However, in so far as the people of any given civil society becomes sophisticated enough to be aware that these national philosophies differ from place to place and also alter over the course of history in any given place, a more sound basis is required to justify one's own customs.[52] This cannot simply be a wise legislator who codified these customs at some remote, and possibly no-longer relevant, time in antiquity.[53]

---

49 Ibid., 206.
50 Ibid., 289–290.
51 Ibid., 257.
52 Ibid., 274.
53 Ibid., 287–288.

Thus Rousseau's second source for the production of the general will is a civil religion that hallows the edicts of a legislator or the customs of ancestors as divinely inspired or mandated.[54] In other words, so far from being opposed to an official religion, according to Rousseau "democracy" presupposes an official establishment of religion, and moreover, one whose *raison d'etre* is to produce eternally justified and unimpeachable norms that will act as a *general will* guiding the legislative action of citizens. Everyone must sincerely attempt to be interpreting this commonly acknowledged civil religion. Rousseau believes that it is absolutely essential that every citizen outwardly conform to "the sacred dogmas authorized by the laws."[55] This is, of course, completely consistent with the original classical Greek democratic tradition, as is Rousseau's restriction of citizenship and its attendant voting rights to adult *males* alone.

Rousseau recognizes that any so-called 'arguments' marshaled to defend a civic religion will not withstand rigorous philosophical criticism. He acknowledges that even the most sound religion that can fulfill the function of eternally justifying certain laws or ethical norms will be irrational.[56] Rousseau can admit this and still advocate such a religion because he believes that the rational quest for truth by pure Philosophy (as opposed to 'national philosophy') and Science do not lead to an understanding of some ultimately transcendent Truth, so much as to a dangerously nihilistic subversion of 'truths' necessary for the functioning of society. Rational inquiry alone cannot provide any ethical or moral content to man in place of his ancestral customs or religious beliefs, it can only foster a perversely faithless skepticism.[57] This being the case, scientists must be limited to practical endeavors that further the aims of the nation and theoretical Science must be secreted away from the common man.[58] Thus the conservative character of Democracy at its origin in classical Greece is not contingent, but inherent to the nature of Democracy.

---

54   Ibid., 287, 289.

55   Ibid., 257.

56   Ibid., 289.

57   Ibid., 258.

58   Ibid., 260.

Philosophy is universalist and so it essentially threatens the patriotism of a democratic society, which is necessarily a unique and closed society on guard against others with which it has irreconcilable differences. Various democracies are inherently culturally relative. Rousseau believes that Philosophers or theoretical scientists must outwardly conform to the religion of the state and the customs of the nation, and under no circumstances should they attempt to 'enlighten' the public. They must adhere to a 'national philosophy' or *ethos*. If they pose a threat to public faith in the religious foundations of society they cannot be tolerated. Any philosopher worthy of the name will pose such a threat because he must follow his "own genius" uncompromisingly, which means without the hypocrisy of outward conformity to something ignorant, and this makes him the enemy of a democratic society.[59] Whereas Mill holds up the Socratic life as the standard of Liberalism, the classical Athenian attitude that led to the murder of Socrates by majority vote sets the precedent for Rousseau's theory of Democracy.

Rousseau consciously tried to restore a sacred unity of the city that he saw to be compromised in postclassical times by Christianity's artificial dualism of temporal power vs. spiritual power that fosters a divided allegiance between the earthly fatherland and the heavenly kingdom.[60] Rousseau was acquainted with Islam, and he saw that Muhammad, ruling in Medina and conquering the Arabian peninsula by the sword, already reunited these two realms to a degree unseen since before the Kingdom of Israel met its demise at the hands of Rome. In the eighth chapter of *On the Social Contract*, entitled "On Civil Religion", Rousseau praises Muhammad for overcoming Christianity's divide between secular power and spiritual authority:

> Jesus...In separating the theological system from the political system... made the state to cease being united and caused internal divisions that never ceased to agitate Christian peoples... since there has always been a prince and civil laws

---

59 Ibid., 258.

60 Leo Strauss, *Natural Right and History* (Chicago: University of Chicago Press, 1965), 254.

[distinct from the Church], this double power has given rise to *a perpetual jurisdictional conflict that has made all good polity impossible in Christian states, and no one has ever been able to know whether it is the priest or the master whom one is obliged to obey...* The spirit of Christianity has won everything [in Europe]. The sacred cult has always remained or again become independent of the sovereign and without any *necessary* link to the state. *Mohammed had very sound opinions. He tied his political system together very well, and so long as the form of his government subsisted under his successors, the caliphs, this government was utterly unified, and for that reason it was good.* But as the Arabs became prosperous, lettered, polished, soft and cowardly, they were subjugated by barbarians. Then the division between the two powers began again... especially in the sect of Ali [Shiites]; and there are states, such as Persia, where it never ceases to be felt.[61]

Firstly, this passage shows that Islam is an example of the kind of "civic religion" that Rousseau believed should ground and reinforce the general will. Secondly, it is clear that Rousseau recognized the fundamental difference between Christianity and Islam. Christianity is essentially inclined to a separation of Religion and the State, and it deviates from its foundation in so far as these are conflated by accidents of history. On the other hand Islam was a political religion from its very inception, and it is moving away from what Muhammad intended it to be in so far as it allows for the emergence of a competing secular political authority. Third and finally, it is most interesting that Rousseau praises the early Caliphate before it was overpowered by the reassertion of more secular and civilized Persians.

The passages above suggest that Rousseau himself would likely have endorsed the theocratic "Islamic Democracies" of our time. More significantly, if a projected demographic shift does indeed make Islam the religion of a majority of people on the Earth before the end of the 21st century, Rousseau would view Islam as the basis of the general will of a global democracy. Of course, this would mean a world governed by *sharia* law.

Carl Schmitt observed that classical democracies were founded on a correct understanding of the concept of *equality*, which must always presuppose those who are unequal in order to have any concrete meaning.

---

61 Jean-Jacques Rousseau, *The Basic Political Writings*, 221–222.

A Democracy is, therefore, the rule by those who *are* equal to one another and respectfully recognize themselves as accountable to *their own* general will. This is true of democracy both in its original conception by the Greeks, and as resurrected in the modern era by Jean-Jacques Rousseau. There can no more be a universal democracy (as so-called 'Social Democrats' vainly imagine) as there can be a universal religion and religious culture. Neither at its origin, nor in its retrieval by Rousseau, has Democracy ever been "secular". The basis of its general will is the religious understanding of the mob, the mob that murdered Socrates.

In his seminal book *The Concept of the Political*, Carl Schmitt argues against the empty definition of the state as nothing more than "the political status of an organized people in an enclosed territorial unit."[62] In such a definition, the concepts of the "state" and the "political" are used to define each other in a circular and tautological manner.[63] The *political agency* or "state" can only exist if it can compel its citizens to use violence against others *and to sacrifice themselves* in defense of the character or common ideals of their society. The *state* is defined by holding such a monopoly of violence, and it exists only so long as no other social, economic, *or religious* force makes demands on its citizens to go out and kill or be killed for some other cause.[64]

Schmitt's definition of the political is grounded on a Heraclitean ontology of *emergence* through strife that is revived by Friedrich Nietzsche and Martin Heidegger. The first edition of *The Concept of the Political* appeared in 1927, the same year as *Being and Time* was published. Schmitt and Heidegger both joined the Nazi party on the same day in 1933. They stood in line together. Despite the existence on record of only one letter exchanged between these men, there appears to have been a significant intellectual and practical relationship between them. Schmitt collaborated with Heidegger in the "bringing into line" of the university, especially of the Faculty of Law, after Heidegger was appointed Rector-Führer of Freiburg

---

62 Carl Schmitt, *The Concept of the Political* (Chicago: University of Chicago Press, 1996), 19–20.

63 Ibid., 22.

64 Ibid., 48.

University.[65] The greatest opposition to the new measures, which included direct appointment of the Rector by the minister of the Badenland (rather than election by fellow professors), was particularly strong in the Faculty of Law. In 1934, Schmitt used his lectures to argue in favor of granting disciplinary and judicial powers to the Rector, allowing him to exercise an absolute dictatorship at the University. With Schmitt's help, by 1937 Heidegger brought at least three prominent jurists into the NSDAP and outmaneuvered his detractors in Freiburg's Faculty of Law. This cooperation was explicitly requested by Heidegger in his letter of August 22, 1933 to Carl Schmitt thanking him for a dedicated copy of the new, third edition of *The Concept of the Political*. That letter reads as follows:

> Most honored Mr. Schmitt,
>
> I thank you for having sent me your text, which I already know in the second edition and which contains an approach of the greatest importance. I would be most appreciative if I could speak with you about this *viva voce* some day. On the topic of your quotation of Heraclitus, I especially appreciated the fact that you did not forget the *basileus*, which alone gives the whole saying its full content, when it is fully interpreted. For years I have had such an interpretation ready, concerning the concept of truth — *edeixe* and *epoiçse* [reveals and makes], which appear in fragment 53. But I now find myself also in the middle of the *polemos*, and the literary projects must give way. I would only like to say to you today that I am counting very much on your decisive collaboration, when it comes to the entire rebuilding, from the inside, of the Faculty of Law, in its educational and scientific orientation. Here the situation is unfortunately quite hopeless. The gathering of spiritual forces, which should lead up to what is coming, becomes increasingly urgent. I conclude today with my friendliest salutations.
>
> Heil Hitler!
>
> Your Heidegger

The approach or beginning (*Ansatz*) of great importance to which Heidegger refers consists of the deletion of the entire first chapter of *The Concept of the Political* and the first paragraph of its second chapter, so that in the 1933 edition it begins immediately with the discussion of the

---

65 Emmanuel Faye, *Heidegger: The Introduction of Nazism into Philosophy* (New Haven: Yale University Press, 2009), 157–158.

distinction between friend and enemy. There are two more specific changes here that are significant in accounting for Heidegger's positive response to this new edition. The friend-enemy distinction is no longer referred to as "specifically" political, but rather as "authentically [*eigentlich*] political", and the word "existential", already used to describe the enemy as constitutionally *other*, is now italicized and repeated for emphasis: "The enemy is, in a particularly intense sense, *existentially* an other and a stranger, with whom, in an extreme case, *existential* conflicts are possible."[66] Both of these changes are nods to the existential terminology of *Being and Time*, especially its discourse of "authenticity" [*eigentlichkeit*].

Heidegger mentions that he is already familiar with the earlier edition of *The Concept of the Political* and Schmitt has clearly read and been influenced by *Being and Time*. However, the similarities between the thought of the two men can better be understood in terms of a certain interpretation of Heraclitus that they share. Interestingly, the 53rd Fragment of Heraclitus, to which Heidegger makes reference in this letter, is not actually quoted by Schmitt in the 1933 edition of *The Concept of the Political*. It is merely alluded to in a note added to this edition, so that one wonders if Schmitt had quoted the fragment itself in his handwritten dedication of the book to Heidegger. Here is the fragment in Charles Kahn's translation: "War is father of all and king of all; and some he has shown as gods, others men; some he has made slaves, others free."[67] Schmitt's note concerns a 1931 text by Alfred Baumler, entitled *Nietzsche, the Philosopher and the Politician*. This text was a common source to both Schmitt and Heidegger. Heidegger praised Baumler's postscript to his edition of *The Will to Power*. Here is Schmitt's note from the 1933 edition of *The Concept of the Political*:

> A. Baumler interprets Nietzsche's and Heraclitus's concept of combat in a totally agonistic mode. Question: Where do the enemies in Walhalla come from? H. Schaefer, in *Form of the State and Politics* (1932), refers to the "agonistic underlying character" of Greek life; even in bloody confrontations between Greeks and Greeks, combat was only "agon", the opponent only an "antagonist", a player

---

66 Ibid., 159.
67 Charles H. Kahn, *The Art and Thought of Heraclitus* (Cambridge: Cambridge University Press, 1999), 67.

on the other team, an opponent, not an enemy, and consequently the conclusion of the contest was not a peace treaty (*eirçnç*). That did not end until the Peloponnesian War, as the political unity of Greece was shattered. In every in-depth examination of war, the great metaphysical opposition of *agonistic* versus *political* thought emerges. Regarding the most recent times, I would like to mention here the great polemic between Ernst Jünger and Paul Adams…., which hopefully will soon be available in print. There Ernst Jünger represented the agonistic ("Man was not made for peace") while Paul Adams saw the meaning of war in the bringing about of domination, order, and peace.[68]

In order to understand the oblique reference to Heraclitus in this passage, and thereby grasp the significance of Heidegger's approval of it in his letter to Schmitt, we must turn to the relevant section of the text of Baumler, where he writes:

> The assumed preconditions of justice are inequality and struggle. This justice does not rule over the world; it does not rule over the tumult of conflicting parties; it knows neither culpability nor responsibility, nor judicial procedure nor sentencing. It is immanent to struggle. That is why it is not possible in a peaceful world. Justice can exist only where forces are given free reign to measure themselves against one another. Under an absolute authority, in an order of things that recognizes a divine master, in the domain of the *Pax Romana*, there is no longer any justice, because there is no longer any struggle there. The world is petrified in a conventional form. Nietzsche, to the contrary asserts: Justice is regenerated at every moment out of struggle, which is the father of all things. It is conflict that makes the master, master and the slave, slave. Thus speaks Heraclitus of Ephesus. But this is also an originally Germanic conception: in struggle is revealed who is noble and who is not; it is by his innate courage that the master becomes master, and it is by his cowardliness that the slave becomes slave. And it is precisely in this way that eternal justice is expressed; it structures and separates, it creates the order of the world, it is at the origin of every distinction of rank. This is how, out of Nietzsche's nodal German/Greek metaphysics, his great doctrine arose: that there is not one morality, but a master morality and a slave morality.[69]

---

68 Emmanuel Faye, *Heidegger: The Introduction of Nazism into Philosophy*, 162.
69 Ibid., 164.

Heidegger, who was familiar with this text by Baumler, goes even further than Baumler in equating the Heraclitean metaphysical notion of the *polemos* with the ethos of the "master morality". This occurs in his lecture course of 1933–34, entitled "The Essence of Truth." It is significant that the still unpublished original version of "On the Essence of Truth" was *first* delivered during celebrations of the withdrawal of French troops from the Rhineland on the *Heimattag* [Homeland Day] of the Badenland, on July 11, 1930. Indeed, the interpretation of Heraclitus that Heidegger claims to have "for years… had ready" in his letter to Schmitt, is forwarded here, with an entire paragraph under the heading "The maxim of Heraclitus. Strife as the essence of the entity." He offers the following exegesis of the 53rd Fragment of Heraclitus, the one made reference to in the letter to Schmitt:

> With grandeur and simplicity, at the beginning of the fragment, there appears the word *polemos*, war [*Krieg*]. What is so designated is not the external event or the advance of the "military" but rather that which is decisive: to stand up against the enemy. I have translated this as "struggle" to seize the essential; but it is also important to reflect on the following: it is not a question of *agon*, vying/competition, in which two friendly [*freundliche*] adversaries measure forces against one another, but of *polemos*, war [*Krieg*]; which means that there is something serious in struggle; the adversary is not a partner, but an enemy. Struggle as a holding out in the face of the enemy; more precisely, as endurance in confrontation.[70]

Schmitt believes that all genuine political theories more or less consider man to be "evil", or as he puts it: "by no means an unproblematic, but a dangerous and dynamic being."[71] The Jewish German political philosopher Leo Strauss, who was a student of Carl Schmitt before immigrating to the United States where he became a tremendously influential professor at the University of Chicago, wrote a critical commentary on the *Concept of the Political*. Schmitt thought so highly of this commentary that he revised the

---

70 Ibid., 167.

71 Carl Schmitt, *The Concept of the Political*, 61.

book in accordance with it, and Strauss' commentary appears appended to revised editions of Schmitt's key text.

In what is perhaps the most important passage in his commentary, Strauss draws the conclusion that Schmitt's argument basically ends in the claim that either Authoritarianism is valid or Anarchism is, but not the mediocre modern liberalism that lies somewhere between these extremes.[72] According to Schmitt such liberalism is a purely negative political philosophy, that is, one that has nothing positive of its own to offer. Believing human beings to be essentially good, liberals promulgate distrust of government and advocate strict limits on its powers.[73] This would only be valid if there were no limit to how undangerous human beings could be rendered through a civilizing education.

According to Strauss, Schmitt would view Anarchism as consistent in so far as it made this basic claim — namely, that it is possible through the perfection of education to eventually render man an entirely undangerous creature busy with entertaining himself without causing any harm to others.[74] The validity of Schmitt's concept of the political lies in denying this anarchist claim. Moreover, this means that the kind of dangerousness or "evil" in man that Schmitt believes cannot be overcome is not merely the innocent 'evil' of a brute beast driven by passions.[75] Potentially, this weak form of evil could be tamed by education in so far as it is a condition of lack, of ignorance, of the want of knowledge and discipline.

Rather than this negative and naïve 'evil', Strauss believes Schmitt to be affirming a positive moral evil in man — one that no measure of education can overcome, and one that greater knowledge is only likely to render more cunningly lethal.[76] Consequently, there is a need for dominion — and this is the ground of authoritarian politics.[77] If the only

---

72 Ibid., 98.

73 Ibid., 60–61.

74 Ibid., 99.

75 Ibid., 99–100.

76 Ibid., 101, 103.

77 Ibid., 100.

consistently liberal 'politics' would be a stateless anarchism then it is no wonder that in practice we never see wholly liberal ideals realized, but only authoritarian regimes moderated by liberal critiques.[78]

Schmitt argues that laws acting as the original foundation of a state must emerge from decisive action required by a concrete existential crisis.[79] This "state of exception" or "state of emergency" is a mirror wherein one sees for the first time who is really in charge and what is actually the fundamental character that governs a given society. The exception is, for Schmitt, "more interesting" than the rule because it is only in reference to the exception that genuine rules for the normal function of a society may be extrapolated without hypocrisy.[80] To put it bluntly, the claim that *under no circumstances do we in our society do such and such, or allow such and such* (for example, torture) can only be sincerely elaborated in law based on its having held true under those actual circumstances.

Schmitt defines the "sovereign" as not only the one who decides whether or when the state of exception exists, but also as the agency that determines what to do under those circumstances.[81] It is important to note that in this sense the word "sovereign" is by no means synonymous with a 'king' or even a single President or Chancellor. A Council or Parliament may act as "the sovereign" so long as its many members are capable of coordinating their actions in the emergency situation as if they were the limbs of one body governed by a single mind. It is backwards for a constitution to define the sovereign's range of acceptable actions in any given state of exception. If a situation's dynamics could be anticipated or described in sufficient detail before hand, so that an appropriate course of action could be formulated in constitutional law, then it would not be a genuine state of *exception*. Rather, it is the sovereign's actions in addressing a dynamic and dangerously unpredictable situation in such a way that this state of exception *is not equivalent to a state of chaos* that first reveals

---

78 Ibid., 70.

79 Ibid., 48–49.

80 Ibid., 15.

81 Ibid., 5, 7.

the actual political *constitution*, or make-up, of a state or catalyzes the *emergence* of a new constitution.[82]

This amounts to a division of the two components of the idea of "legal order" in such a way as the latter of the two concepts (order) outlives the former (literal law) and either ultimately reaffirms or redefines it.[83] The military chain of command, which is brought into effect as the constitution is suspended in martial law, is an example of this.[84] While the law is abrogated or a new law is being formed, there is still a juristic state of order, wherein the sovereign, as commander-in-chief issues orders down a hierarchical chain of command.[85] These are to be obeyed by subjects, who offer themselves in this subjection in exchange for the sovereign's offer of protection of their private wills and his promise of restoration of an order wherein they may be exercised.[86] Only the sovereign decides when the state of emergency has come to an end. In this event, either the old constitution is restored by the sovereign, or the sovereign literally re-constitutes the state.

Schmitt believes that this 'miraculous' quality of the authentic and effective exercise of sovereignty reveals the fact that the concept of state sovereignty is a substitution for the theological conception of the sovereignty of God. In *Political Theology* he observes that, by analogy, the sovereign departs from the normal legal order of the state in the way that a miracle or 'act of God' departs from the normal divine order of the universe.[87] This would seem to suggest that the actions of a sovereign in an emergency situation must either embody the deepest religious convictions of his subjects, or they must in effect establish new and even deeper convictions that supplant the established religion of the society in crisis. Thus a political theory is incoherent if it conceives of sovereignty as in some way

---

82 Ibid., 6–7, 14.

83 Ibid., 12.

84 Ibid., 28.

85 Ibid., 12–13.

86 Ibid., 52.

87 Schmitt, 1996: 42; Carl Schmitt, *Political Theology* (Chicago: University of Chicago Press, 2006), 36.

divided between the spiritual ideals of a people and some different ideal motivating the state's rule of law, such as implementation of the standards of the *Universal Declaration of Human Rights*.

Schmitt argues that "Humanity" is not a coherent concept on the basis of which to elaborate a theory of political Right, that is — unless humanity as a whole were threatened by a non-human, presumably extraterrestrial, enemy so alien that in respect to it "we" recognized that we do share a common way of life that we must collectively defend against "them".[88] Otherwise: "Humanity is not a political concept, and no political entity or society and no status corresponds to it."[89] Any political use of the term is polemical and propagandistic. A universal human *state* is incoherent because it must presume the continued existence of something 'inhuman'.[90] This raises another key aspect of Schmitt's thought. That is, laws that act as the foundation of a state should not be determined by means of the purely ideal or theoretical reflection of intellectuals. They must emerge from decisive action required by a concrete existential situation, namely the existence of a real enemy that poses a genuine threat to one's way of life.[91] This is, according to Schmitt, the fundamental reason why a world state is not possible:

> The political entity presupposes the real existence of an enemy and therefore coexistence with another political entity. As long as a state exists, there will thus always be in the world more than just one state. A world state which embraces the entire globe and all of humanity cannot exist. The political world is a pluriverse, not a universe. In this sense every theory of state is pluralistic, even though in a different way from the domestic theory of pluralism... The political entity cannot by its very nature be universal in the sense of embracing all of humanity and the entire world... Humanity as such cannot wage war because it has no enemy, *at least not on this planet*. The concept of humanity excludes the concept

---

88 Carl Schmitt, *The Concept of the Political*, 54.

89 Ibid., 55.

90 Ibid.

91 Ibid., 48–49.

of the enemy, because the enemy does not cease to be a human being—and hence there is no specific differentiation in that concept.[92]

In his *Theory of the Partisan*, written some four decades later, Schmitt's contemplation of the transformative power of technology has brought him to reconsider this position on the coherence of a world state whose sovereign power is constituted by, and endures through, a concrete emergency. Perhaps it is not so much a reconsideration of his position as an elaboration at length on that ominous qualifier in the passage above: "at least not on this planet." Some believe that technical progress will eventually render the partisan and his grievances as irrelevant as the agrarian revolution did to the Stone Age hunter.[93] However, Schmitt counters this techno-utopian view with a techno-dystopian scenario wherein the partisan radically adapts to technological conditions in two related ways.[94]

First, there is a dramatic transformation in *space*. He acknowledges that increased technical-industrial progress could abolish the distinction between the land-rooted partisan and the sea-faring pirate.[95] We saw a preview of this with German introduction of the submarine into conventional sea-surface warfare. At first, it was condemned by England as a criminal activity fit only for pirates, but eventually it came to be recognized as a brute fact of technological progress that radically transformed the ethic of sea warfare.[96] This potential is especially great when one thinks of the colonization of space, which Schmitt views as a vast new lawless arena for competition in conquest—one that encompasses the Earth and calls into question its entire *nomos*.[97] Schmitt writes:

> Thus, our problem is widened to planetary dimensions. It even reaches further to supra-planetary dimensions. Technical-industrial progress makes possible the journey into cosmic spaces, and thereby opens up equally immeasurable

---

92 Ibid., 53–54.

93 Carl Schmitt, *Theory of the Partisan* (New York: Telos Press, 2007), 77–78.

94 Ibid., 78–79.

95 Ibid., 70–71.

96 Ibid., 71–72.

97 Ibid., 80.

new possibilities for political conquests, because the new spaces can and must be appropriated by men. Old-style land and sea appropriations, as known in the previous history of mankind, will be followed by new-style space appropriations. *Appropriation* will be followed by *distribution* and *production*. In this respect, despite all further progress, things will remain as before. Technical-industrial progress will create only a new intensity of new appropriations, distributions, and productions, and thereby only intensify the old questions. ...Consequently, these immeasurable spaces also become potential battlefields, and the domination of this earth hangs in the balance. The famous astronauts and cosmonauts, who formerly were only propaganda stars of the mass media (press, radio, and television), will have the opportunity to become cosmopirates, even perhaps to morph into cosmopartisans.[98]

Second, there is the increasing deadliness of weapons of mass destruction.[99] From his point of view, late into the Cold War but before the end of it, Schmitt saw half of humanity held hostage by partisan-governments that control the other half, which would have to be totally dehumanized before it was incinerated.[100] Yet, even beyond this, the miniaturization of these weapons could bring us to a point where partisan warfare with tactical nuclear weapons becomes a possibility.[101] Schmitt clearly and distinctly foresaw the threat of nuclear terrorism in an age when the terrorist was politically decisive for the lives of would-be citizens, not the sovereign authorities of their respective governments.[102]

Combine these two technical developments, namely space colonization and the proliferation and miniaturization of weapons of mass destruction, and the elemental indestructibility of Earth itself is called into question, let alone the significance of the patch of native soil defended by the partisan of old. We have in interplanetary conflict a threat to Earth as a whole, which according to the logic of Schmitt's own argument ought to justify a world sovereign. This is even more true if we substitute his

---

98 Ibid.
99 Ibid., 94.
100 Ibid., 93–94.
101 Ibid., 76.
102 Ibid., 73.

technological catalyst with the specter of convergent advancements in technology tending towards a technological singularity, innovations that do not represent merely incremental or quantitative change but qualitatively call into question the human form of life as we know it. This singularity would then have to be conceived of, in political terms, as a *world state of emergency* in two senses: a *state of emergency* of global scope, and a *world state* whose constitutional order emerges from out of the sovereign decision made therein. That is the case I will make over the next several chapters, with a view to specific and convergent technological innovations, before returning to consider just what kind of World State can secure us a human, rather than inhuman, future beyond the planetary emergency of a technological apocalypse.

CHAPTER 3

The Neo-Eugenic World State

The emergence of biotechnologies with the potential to yield either a positively superhuman or a horrifically inhuman future represents a serious technological catalyst for the establishment of a world state. In the half-century since the inauguration of the atomic age, just as Physics has begun to yield its place as the paradigm science for technological innovation to Biology — so also the subtler dangers of unreflectively manipulating *life* have largely supplanted the specter of mass death amidst a worldwide nuclear conflagration. The advent of human biotechnologies such as cloning, germ-line genetic engineering, and pre-implantation embryonic selection, confronts us with the prospect of a new Eugenics. The New Eugenics of human biotechnology is *now* what nuclear physics was in the 1930s, when it became clear that the nation first to build atomic bombs could bend the world to its will. Civilizational values matter, and we ought to consider very seriously whether it should be the collectivist Chinese — with their instrumentalist and conformist mentality — who become the first, and for a time possibly the only, nation to implement the New Eugenics.

So far as we can tell the founder of Eugenics theory was the Greek philosopher Plato, an element of whose utopian society in *The Republic* (written circa 380 BC) is a selective breeding program aimed at the maintenance of a caste system dominated by an intellectual aristocracy.[1] From 459a to 461e in *The Republic*, Plato argues that techniques used from ancient times

---

1   Richard Lynn, *Eugenics: A Reassessment* (Westport: Praeger, 2001), 3, 9.

by farmers to produce better livestock, or thoroughbred hunting dogs or horses, or by agriculturalists or botanists to cross-pollinate flowers and cultivate new kinds of plants, can be applied to improving the genetic stock of human beings as well.[2] Plato thinks that the ideal state ought to set in place a selective breeding program that will encourage the most genetically fit males and females to breed most frequently and discourage breeding among the most genetically unfit males and females. These Eugenic policies should also incorporate population control mechanisms aimed at keeping the number of people in the society constant as scaled to its pool of resources. Deformed children or those with readily identifiable hereditary diseases would be sequestered from the rest of the population. Any fetuses conceived through marriages unauthorized by the state, in violation of Eugenic or population control policies must be aborted or, if born, should not expect state support.

In even more remote antiquity than Plato's era, plants had been improved by artificial selection (the first sweet strawberries are a product of such methods). Later, during the Middle Ages, exceptionally sturdy horses were bred to bear Knights and their armor; eighteenth century English stockbreeders produced improved strains of cattle and sheep; and racehorses are all products of selective breeding.[3] The application of such methods *to human populations* figured in several other utopian texts throughout the centuries, but it was not worked out in theory and practice until the late 19th century.

It was then that, in a series of publications from the 1860s through the 1880s, the English statistician, biologist, and psychologist Sir Francis Galton elaborated the concept of *eugenics*—derived from the Greek for "good breeding"—as a methodology for the improvement of the genetic qualities of distinct human populations with respect to health, intelligence, and character/disposition.[4] Galton argued that these three qualities are vital to the progress of civilization and he aggregated them into a single

---

2   Plato, *The Republic* (Basic Books, 1991), 138–140.

3   Lynn, *Eugenics: A Reassessment*, 9.

4   Ibid., 4.

criterion of human "civic worthiness", arguing that it is in the interests of the State to check barbarism and promote scientific, cultural, and economic achievement by cultivating the genetic "worth" of its population.[5]

At a time when many of his contemporaries argued that intelligence was contingent on environmental factors alone, Galton was one of the first scientists to postulate a factor of heritable general intelligence and marshal empirical evidence in its favor.[6] One of his most impressive arguments was based on studies of gifted orphans and foundlings, such as the mathematician Jean D'Alembert, who were raised by ordinary or downtrodden parents that were not their progenitors.[7] He also studied twins who were reared apart from one another and found that the striking similarities between them attested to the fact that the influence of heredity on the general traits of intelligence, health, and character was stronger than that of environment.[8]

Galton did not conceive of Eugenics as an enhancement measure so much as a corrective one. Under the influence of Charles Darwin (who in turn was influenced by Galton into accepting Eugenics), Galton developed the idea that the force of natural selection of the fittest decreases as a given civilization advances. In the early stages of a civilization, he argued, it is the most able, intelligent and enterprising individuals who have the most children. Often this involved a small number of the best males breeding with a larger number of the better females — as in the Near Eastern, Indian, and Chinese civilizations.[9] However, as a civilization grows more refined the elite of men *and women* have fewer children on account of their being preoccupied with advancing their careers. They marry later, or not at all. Consequently, the underclass of the civilization breeds at a far more prodigious rate than the elite that first raised a given population out of barbarism and onto the world-historical stage, so that

---

5   Ibid., 4–5.

6   Ibid., 5.

7   Ibid., 6.

8   Ibid., 6–7.

9   Ibid., 9.

the civilization in question enters a phase of decadence and terminal decline.[10] Galton viewed the artificial selection of Eugenics as a means to counter this *dysgenic* process.

Finally, Galton formulated certain basic proposals for the implementation of Eugenics that were broadly adhered to by various organizations and programs in the 20th century. To start with he conceived of a tripartite division of society into "desirables", "passable" persons, and unquestionably "undesirable" elements. The majority of the population would fall into the "passable" class and would not be the object of any eugenic intervention. This was thought to decrease the likelihood of their opposition to positive eugenic measures aimed at the "desirables" and negative ones directed at "undesirables." The "desirables" would be bound together through social institutions inculcating them with a special sense of responsibility and provided with financial incentives to procreate with others of their own kind.[11] Exceptional individuals from lower social classes would be given scholarships allowing them to rise into the elite.[12] Sterilization would be mandated for the "undesirables" — consisting of the mentally retarded, the deformed, habitual criminals, substance abusers, and others with low intelligence, hereditary health issues, and weak character.[13] To safeguard the gene pool of the nation, dysgenic immigration of undesirables into the eugenic state would be checked while their emigration to elsewhere would be encouraged.[14] Galton was aware that such policies could not be implemented democratically, and he argued that a eugenic state needs an authoritarian constitution (like present day China).[15]

One of the first converts to Galton's theory of Eugenics was Charles Darwin, who wrote to Galton of being convinced by his arguments and whose 1871 book *Descent of Man* remarks affirmatively on Galton's

---

10   Ibid., 7–8.

11   Ibid., 10.

12   Ibid., 13.

13   Ibid., 11–12.

14   Ibid., 12–13.

15   Ibid., 16.

hypothesis of dysgenic decline of civilizations.[16] Herbert Spencer was also an early proponent of Galton's views. By the early 1920s and onward through the first half of the 20th century, views favorable to Eugenics prevailed among members of the scientific, intellectual, literary, educational, juridical and political elite of Europe and America, among whom were numerous Nobel Prize winners and leading political figures from across the ideological spectrum from liberal to authoritarian and capitalist to socialist.[17] In Europe, the advocates of Eugenics included: Sir Ronald Fisher, Sir Julian Huxley, Sir Peter Medawar, J.B.S. Haldane, Francis Crick, Charles Spearman, Alexis Carrell, Bertrand Russell, Sydney Webb, George Bernard Shaw, Arthur Balfour, Winston Churchill, Maynard Keynes, Sir William Beveridge, and H.G. Wells. Those in America who found Eugenics a sound and practicable theory were just as eminent and influential, among them: Charles Davenport, Hermann Muller, Linus Pauling, Joshua Lederberg, Edward Thorndike, William McDougall, Charles Wilson, Irving Fisher, Oliver Wendell Holmes, Margaret Sanger, and Theodore Roosevelt.

The governments of the United States (at the level of individual states), Canada, Denmark, Switzerland, Norway, Sweden, and Finland all officially implemented eugenic policies, including strict immigration restriction laws and hundreds of thousands of forced sterilizations of the hereditarily diseased, the mentally disabled, and habitual criminals.[18] Additionally, numerous mainstream European and American non-governmental organizations were established for the promotion of eugenics policies, such as: the Eugenic Education Society (1907, later renamed the British Eugenics Society), the Eugenics Records Office (1910) at Cold Spring Harbor, the American Genetics Association (1913), the American Eugenics Society (1923), and special programs at the Galton Laboratory in London and at Cornell Medical School in New York endowed by the Carnegie Institution

---

16 Ibid., 19.

17 Ibid., 20–27, 293–294.

18 Ibid., 34–35.

and the Rockefeller Foundation in the 1930s.[19] Some of these private organizations focused on those policies that were considered impracticable for government programs, such as birth control advocacy. The American feminist Margaret Sanger, who was the leading pioneer in the campaign for the legalization and provision of birth control, was an ardent eugenicist.[20]

In light of the above, we can see that the now prevalent association of Eugenics theory and practice with Nazi German pseudoscience and genocide is inaccurate. The "genetic hygiene" program that began at the Kaiser Wilhelm Institute in the 1920s was just one among many similar institutions throughout the Western world. Moreover, the sterilizations carried out by the German government during the Nazi period — under laws explicitly modeled on those already in place in the United States — were primarily aimed at freeing up medical resources (hospital beds, trained staff, etc.) under wartime conditions, and they pale in comparison to the scale of sterilizations in, for example, Sweden.[21] Most importantly, there is no good evidence that Nazi persecution of the Jews had anything to do with "eugenic" ideas of their racial inferiority. In his writings on curbing dysgenic immigration Francis Galton approved of Jewish immigrants from Eastern Europe and Russia on account of their relatively high genetic worth — especially with respect to general intelligence.[22] Likewise, if the Nazis persecuted the Jews on a genetic basis it would be because "this little nation" was deemed "the mightiest counterpart to the Aryan."[23] This was Adolf Hitler's own view, as well as that of Josef Mengele. Furthermore, scientific eugenicists who do argue for racial genetic inequalities claim that it is East Asian (not Aryan) populations that lead the world in general intelligence, and they do not believe that race mixing is inherently dysgenic.[24]

---

19 Ibid., 24, 27.
20 Ibid., 33.
21 Ibid., 28.
22 Ibid., 13.
23 Ibid., 29.
24 Ibid., 29.

In the first generation after the Second World War, there was a precipitous decline in the acceptance of Eugenics in theory, let alone in its application. This branding of Eugenics as "pseudoscience" would have been totally unpredictable in the early to mid 20th century and it is more than likely attributable to a single factor: the mistaken conflation of Eugenics with wartime Nazi horrors — the full extent of which were not psychologically processed by the public until the 1960s. In both Europe and America, governmental eugenic policies such as sterilization and restricted immigration were ceased, the major Eugenics organizations all changed their names and those of their journals (usually to something involving "social biology"), and a vast array of anti-Eugenic civil liberties groups arose.[25]

On the Heideggerian definition of "technology", which reaches back to the meaning of the Greek concept *techne* ("craft, technique, means of cultivation"), classical Eugenics methods are already a human *biotechnology*. However, what is usually meant by *human biotechnology* is a number of more recent developments such as: pre-implantation embryonic selection, cloning, and genetic engineering. As early as 1969, Robert Sinsheimer coined the term "the new eugenics" to describe the potential uses of these technologies to improve human genetic quality more effectively and less controversially than selective breeding, sterilization, and other classical methods.[26] Many medical professionals — who claim to be solely concerned for the health of their patients — have denied that the new human biotechnologies are eugenic. Nevertheless, since their impact is not confined to individual patients but affects society as a whole, and because this effect is one that improves the collective gene pool, such technologies *are inherently eugenic*.[27]

On this point, the critics have a clearer view than some of the proponents of genetic enhancement uses of the new human biotechnologies, who are unwilling to call a spade a spade. Opponents of genetic

---

25 Ibid., 37.

26 Ibid., 243, 270–271.

27 Ibid., 256.

enhancement have consistently labeled the new human biotechnologies as "eugenic."[28] These include at least two members of the Presidential Council on Bio-Ethics—the Chairman Leon Kass, who wants a ban on genetic engineering, and Gilbert Meilander who identified the arrival of biotechnological genetic enhancement as the advent of "a new era of eugenics" which he compared to Nazi atrocities. While on visit to the Holocaust Museum in Washington D.C., Boston University bioethicist George Annas also went even further, to make the claim that "Modern genetics [in general] is eugenics." Annas has proposed a United Nations ban on genetic engineering. Hysterically hostile reactions such as these are completely insensitive to the differences between various emergent biotechnologies. Genetic engineering, cloning, and embryonic selection on the basis of pre-implantation genetic diagnosis, should not all be haphazardly conflated and consummately rejected as equally threatening to what we take to be a meaningful human life. We ought to carefully examine the respective promises and perils of each human biotechnology.

The most controversial of the new eugenic biotechnologies is human *cloning*. The term is derived from the Greek word *klon*, meaning "twig," which refers to an ancient method for the cultivation of identical copies of fruit trees by means of planting cuttings or twigs from the original tree.[29] Cloning, in general, is the reproduction of an identical copy of a plant or animal. In the 1950s a certain frog became the first animal to be cloned, while the first mammal, Dolly the Sheep, was cloned in 1997; shortly thereafter 500 identical copies of cattle embryos were cloned, and in 1998–1999 cloned mice were being produced.[30] The methodologies for the cloning of mammals that could be adapted for use on humans, consist of the following three techniques:[31] 1) A cell from a mature animal can be reverted to an undifferentiated embryo and implanted into the host mother before re-differentiation into the various bodily organs [This

---

28 Ramez Naam, *More Than Human* (New York: Random House, 2005), 143.
29 Ibid., 255.
30 Ibid.
31 Ibid.

was the method used in the production of Dolly the Sheep]; 2) nuclei of the cells of an unfertilized egg or an early embryo could be removed and replaced with those taken from another embryo (or a mature adult, as in 1); 3) embryos could be grown in vitro before being split into two or more embryos, as happens naturally in the prenatal development of identical twins. This third technique could be used to produce multiple clones at once. It is also possible to assess the genetic characteristics of one of the embryos before deciding whether the other identical ones should be implanted. On December 15, 1998, researchers in South Korea used this technique to successfully clone a human embryo that was allowed to develop into a blastocyst of four cells before being destroyed.[32]

The possibility of using cloning technologies to produce (nearly) identical copies of humans has been widely condemned. The spirit of opposition to human cloning *as a eugenic technology* is perhaps epitomized by a resolution condemning it that was passed by the European Parliament in 1997: "In the clear conviction that the cloning of human beings... cannot under any circumstances be justified or tolerated by any society, because it is a serious violation of fundamental rights and is contrary to the principle of equality of human beings as it permits a eugenic and racist selection of the human race, it offends against human dignity and it requires experimentation on humans... Each individual has a right to his or her own genetic identity, and human cloning is, and must continue to be, prohibited."[33]

That a cloned child would resemble only one parent, and would have to be raised as a son or daughter by the parent of the opposite sex, would raise troubling questions when the clone reached an age of sexual maturity.[34] A clone is explicitly modeled *on someone else*, whether a parent, a dead sibling, or even worse, a dead ancestor or a celebrity whose genetic code has been sold on the free market. It is often the case that parents who are favorable to the idea of cloning a deceased child emotionally feel

---

32   Ibid.

33   Ibid., 268.

34   Francis Fukuyama, *Our Posthuman Future* (New York: Picador, 2002), 207.

as if it is a way to retrieve a person who was lost to them, even if on an intellectual level they understand that the clone will be a different person.[35] Depending on how old the dead child was this could be very frustrating for the clone, since he may repeatedly fail to fulfill the expectations of the parent. It could also be unfair to the lost child, whose memory might be supplanted by a 'better copy' of him.

Perhaps in part due to the controversies associated with the germline genetic engineering and cloning of humans, embryonic selection by means of Pre-implantation Genetic Diagnosis (PGD) is the most promising of all of the technologies of the New Eugenics. Counseled abortions subsequent to prenatal diagnosis have already significantly decreased the birth incidence of well-identified conditions such as anencephaly, spina bifida, hydrocephaly, Tay-Sachs disease, and B-thalassemia. In the late 1980s the Royal College of Physicians in Britain made an assessment that if all existing procedures to limit inheritance of genetic disorders were effectively implemented, the birth incidence of such disorders could be cut in half.[36] Pre-implantation embryonic selection is on a continuum with prenatal diagnosis, but it is much less stressful for prospective mothers than post-screening abortion (which still leaves the possibility of another disordered fetus in a second attempt) and potentially far more effective in improving the genetic quality of future generations.[37]

PGD was initially developed as a way to make in vitro fertilization (IVF) more effective. Embryos are grown in vitro to the eight-cell stage, at which point one or two of the cells of a given embryo is removed and tested for its genetic composition. The basic procedure of embryonic selection consists of hormonally stimulating a woman to ovulate a number of eggs that are fertilized in vitro.[38] Embryos are then grown in vitro to the eight-cell stage, at which point one or two of the cells of a given embryo is removed for analysis of its genetic composition. At this point in

---

35 Naam, *More Than Human*, 139.

36 Lynn, *Eugenics: A Reassessment*, 280.

37 Ibid., 252.

38 Ibid., 252, 284.

embryological development any one of the eight cells could go on to form the various bodily organs, which is what the remaining six or seven cells do in the embryo chosen as the best for implantation.[39] The first trials of the technology were carried out in the late 1980s and by 1995 there were 16 centers in various countries offering the service. At first it was used only to screen out a variety of readily identified genetic diseases and disorders, but prospective parents are now also using it to select the sex of their child.[40]

If embryonic selection is to move beyond this stage and fulfill its potential as a eugenic technology, three major developments have to take place. First, at present embryonic selection only has a negative eugenic function, and at that a limited one. The genes for a number of late onset diseases and disorders remain to be discovered. If this technology is to serve a positive eugenic function, it will be necessary to identify the genes for general intelligence and special cognitive abilities such as mathematical, linguistic, and musical aptitude, as well as those for physical build and appearance.[41] Second, as it stands now, the number of eggs that a woman can be hormonally stimulated to produce is limited, especially as compared to the potential store of tens or hundreds of thousands in her ovaries. Recently there has been success in raising the number from around a dozen to 25, and it is projected that this trend will continue to the point where a couple has hundreds of embryos to choose from so that they are not forced to make difficult decisions that compromise on one genetic quality versus another.[42] Third, and finally, at present only about half of selected implanted fetuses lead to successful pregnancies. This success rate would have to be increased close to 100%. Alternatively, the optimal embryo could be cloned before implantation, so that it would be available for a second or third attempt in case the first one fails.[43]

---

39 Naam, *More Than Human*, 113–114.
40 Lynn, *Eugenics: A Reassessment*, 252.
41 Ibid., 283.
42 Ibid., 284.
43 Ibid.

As solutions continue to emerge for these problems, embryonic selection will become the most promising of the new eugenic technologies because it allows couples with even slightly below average genes to produce above average offspring through selection of only those fetuses at the high end of the potential range for that couple. For example, with respect to general intelligence, the range between the highest and lowest IQ possible for fetuses produced by a given couple is around 30 points — so it is possible for parents with IQs of only around 90 to produce a child with an IQ of 105.[44] In terms of physical build and appearance, embryonic selection among hundreds of potential fetuses would allow a couple to draw on all kinds of recessive traits that would have a freakishly low probability of phenotypic expression through random procreation. One also avoids the risk of removing a gene from the germ-line that may later be found to have had other, unknown, positive functions.

Genetic engineering was the collaborative invention of Stanley Cohen and Herbert Boyer. Cohen was a professor at Stanford University working on plasmids — small rings of DNA that carry the genes of bacteria. Boyer was a researcher at UC San Francisco who discovered restriction enzymes — molecular tools that could be used for splicing strands of DNA. In 1973, Boyer and Cohen used restriction enzymes to isolate a specific gene and deliver it into *e. coli* bacterium.[45] Subsequently, researchers successfully explored the use of viruses as vectors for the modification of genes in other organisms. Viruses are incapable of reproducing themselves. They infiltrate a cell and reprogram it to produce more of the virus one protein at a time. This process is repeated exponentially, thereby propagating the virus throughout the given organism. Building on the work of Boyer and Cohen, researchers found a way to reprogram a virus so that the cells it 'infects' are given instructions to produce more of a certain gene.[46] Such a method was first employed for the genetic modification of foods — for example, tomatoes that can survive freezing, or pesticide resistant corn.

---

44 Ibid.

45 Naam, *More Than Human*, 14.

46 Ibid., 15.

By the end of the 1980s, French Anderson and his colleagues suggested that the same procedure could be used as a therapy to improve the health of human beings lacking one vital gene or another.[47]

As one might imagine, using viruses that are meant to propagate diseases throughout the body as vectors for genetic engineering does come with some complications. One problem is that the virus might trigger an immune response that could prove damaging or even deadly to the subject.[48] Another problem is that virus vectors are blunt instruments. They insert themselves just about anywhere in the genome targeted for alteration, and while for the most part this does not cause any difficulty (because of the large percentage of non-coding sequences in DNA), it might happen that the vector inserts itself right in the middle of some important coding sequence — such as one intended to fight tumors.[49] Fortunately, a single solution to both of these problems has been discovered.[50] About 80% of the world population is already infected with adeno-associated virus or AAV, which does not seem to cause any disease. This means that selecting the AAV virus as the vector for genetic engineering can bypass a potential immune response in most subjects. AAV is also known to insert itself at one very particular part of the genome, far away from any important coding sequences. In 2002 a University of Washington group led by Roli Hirata and David Russell used this technique to replace a precisely targeted genetic sequence.[51]

There are two types of virus vectoring used in genetic engineering.[52] *Non-insertional vectors* deliver the genetic information into the cell for replication, but without penetrating the cell nucleus. This makes the inserted genetic information vulnerable to destructive processes within the cell, so that it breaks up over time, and it is not carried over when the

---

47 Ibid.
48 Ibid., 29.
49 Ibid., 30.
50 Ibid., 31.
51 Ibid., 32.
52 Ibid., 19–20.

cell divides. Consequently, non-insertional vectors are used for somatic gene therapies, the effects of which may last weeks or months. They are not permanent in the absence of continued therapy because only somatic cells are affected—those not involved in reproduction.[53] By contrast, *insertional vectors* penetrate all the way into the cell nucleus. All of one's DNA is encoded on the 23 pairs of chromosomes protected from damage behind the wall of the nucleus, and the insertional vector splices *this* DNA—which replicates at cell division. This use of insertional vectors is known as germ-line genetic engineering, wherein any single intervention is sufficient to have the new genes passed down to subsequent generations engendered by the engineered subject. Moreover, unlike somatic cell therapy, which is never effective in vectoring new genes to anywhere near *all* of the relevant cells, not only does the subject of germ-line engineering have the new genes in all of his cells, it is also the case that his heirs have at least a 50% chance of being born with the new genes already present in *all* of *their* cells.[54]

In other words, germ-line modification is easier and also more effective than somatic therapy. Somatic-line gene therapy is the only one of the emerging human biotechnologies that is not eugenic, and that could even be considered *dysgenic*. This therapeutic modification of somatic cells in adults allows them to pass on the genetic diseases and disorders for which they are being treated down to subsequent generations. In addition to being the only method of removing defective genes once and for all, germ-line genetic engineering is the only childbearing option for women with genetic mitochondrial diseases, which are passed down to all children.[55] It is also the only solution in cases where the same recessive disorder is present in both parents.[56] Germ-line genetic engineering holds up the potential for a wide range of non-specializing enhancements, in other words eugenic modifications of the genome that do not involve

---

53 Ibid., 116.

54 Ibid., 117.

55 Lynn, *Eugenics: A Reassessment*, 272.

56 Ibid., 272.

designing a specific individual for a specific purpose and thereby potentially degrading her individuality. Rather, enhancements of strength and physique, cognitive functioning, and a lengthened lifespan are potentially eugenic modifications that could lead to a broad-based, population-wide improvement of the capacity for human flourishing. Let us look at each of these eugenic applications of genetic engineering.

The same treatments now being developed for Amytrophic Lateral Sclerosis (aka. Lou Gehrig's disease) by researchers such as Jeffrey Rothstein at Johns Hopkins could potentially be used to genetically enhance strength.[57] Lee Sweeney at the University of Pennsylvania has found that, even if they remain sedentary, mice bred with extra copies of the insulin-like growth factor (IGF-1) gene develop to be as muscular and strong as unmodified mice put on an intense strength-training program. Both the enhanced mice and the exercised ones were 16% stronger than the control group. When the altered mice are put on the same strength-training program, they develop to be as much as 30% stronger than unmodified mice.[58] Applied to humans, imagine how that would translate into a competitive advantage in certain athletic abilities.

A team led by Geoffrey Goldspink at the University of London found a 20% increase in muscle mass and a 25% increase in strength in mice genetically modified by Mechano Growth Factor (MGF), which is similar to IGF-1. Se-Jin Lee at Johns Hopkins took a different approach to genetic enhancement of strength. He knocked out the muscle-growth inhibiting hormone Myostatin in mice, which led to them developing two to three times their normal muscle mass and it also unexpectedly resulted in their having 70% less body fat. Obesity, which is particularly prevalent in America, increases the risk factor for heart disease, diabetes, and cancer. It could be caused by some forty different mutations of six genes.[59] In light of this fact it is all the more significant that in 1996, researchers at the Howard Hughes Medical Institute at Chevy Chase, Maryland successfully

---

57 Naam, *More Than Human*, 24.

58 Ibid., 25.

59 Ibid., 128.

corrected obesity in mice and rats by injecting them with extra copies of the gene for the production of Leptin, a hormone that regulates bodyweight and controls metabolism. Implementation of population wide genetic enhancement of strength and physique, by making it much easier to boost muscle mass and by eliminating the genetic factors of obesity, would turn being weak and fat into strictly psychological problems of personal motivation.

'Intelligence' is a concept so broad and vaguely defined that it would be misleading to suggest that it could be genetically enhanced. Leadership qualities, the depth of one's understanding of acquired knowledge (as opposed to a command of bare facts), motivation, creativity, emotional disposition and interpersonal skills may all factor into whether or not a person is perceived to be 'intelligent.'[60] There is, as yet, no reason to think that these important but fuzzy factors can be manipulated biotechnologically. However, there are several other facets of cognitive functioning that also factor into 'intelligence' broadly construed. These include the number of healthy and active neurons (nerve cells) in the brain, the endurance of the long-term potentiation (LTP) that develops when neurons fire in synch and reinforce their connections to one another, the strength of the synapses across which neurotransmitters travel from pre-synaptic neurons to receptors, and the production of proteins that allow these receptors to respond to neurotransmitter molecules. There are genes which can affect each of these various processes in the brain, and experiments in animals as diverse as fruit flies, mice, and sea slugs have shown that enhancement techniques involving them could improve memory or learning time as well as problem solving abilities involving mathematical, verbal, and spatial skills.[61]

In 2001, Mark Tuszynski and others at the University of California at San Diego developed a treatment for Alzheimer's based on genetically modifying neurons to have extra copies of the gene for Nerve Growth Factor (NGF), which not only leads to the growth of more neurons and

---

60 Ibid., 133.

61 Ibid., 137–138.

nerve connections in the brain, but also restores the health and activity of failing neurons in old brains.[62] Unsurprisingly, NGF could also be used on a healthy brain for enhancement purposes. Around the same time, Howard Federoff and his colleagues at the University of Rochester used NGF to produce a 60% increase in the memory capacity of mice subjected to maze navigation tests. They also found that mice that were both subjected to the NGF genetic modification *and* raised in a more cognitively challenging environment (enclosures with more toys, puzzles, etc.) did five times better at navigating mazes than genetically unmodified mice *and four times as well as unmodified mice raised in the same challenging environment.*[63]

The most famous cognitively enhanced mice are the 'Doogie Mice' (named after Doogie Houser M.D.) produced by Joe Tsien at Princeton University. Tsien's mice were genetically engineered to have extra copies of a gene for the production of NR2B, a vital part of the NMDA receptor that responds to the glutamate neurotransmitter molecule.[64] The production of NR2B declines with age. While Tsien's reversal of this has been shown to potentially increase the risk of stroke or drug addiction, there are other similar enhancements without side effects.[65] In the 1980s Nobel Prize winner Eric Kandel found that it was possible to build the strength of neuronal synapses by increasing production of the chemical CREB. In 1994 Tim Tully at Cold Spring Harbor Laboratory in New York, built on Kandel's discoveries by modifying the genes of fruit flies and sea slugs to produce more CREB, thereby dramatically boosting their rate of learning.[66] These lines of research and development all suggest genetic enhancement of certain types of cognitive functioning may allow humans who benefit from them to more efficiently study for exams geared towards memorization of facts, to learn more languages and musical instruments

---

62 Ibid., 39.
63 Ibid., 40.
64 Ibid., 40–41.
65 Ibid., 42.
66 Ibid., 42–43.

and to do so more quickly, as well as to have an easier time solving complex technical problems that require mathematical and spatial skills.[67] Studies such as those conducted by Harvard professor Robert Barro have shown a significant correlation between the average IQ of a nation and both its gross domestic product (GDP) and rate of economic growth.[68]

At present age-related diseases are to account for half of all deaths.[69] Genetic research into lengthening lifespan began in 1979 with the work of Tom Johnson at the University of Colorado. Initially, Johnson assumed that the rate of aging was controlled by hundreds of different genes and would consequently be very difficult to affect. However, in 1983 his colleague Mike Klass carried out some studies on mutant strains of nematode worms that suggested otherwise. After Klass left the university for work in the private sector, Johnson farmed out Klass' research on the worms to David Friedman, an undergraduate student. Friedman found that aging in the nematode worms was controlled by just a single gene, dubbed *age-1*, and his work with Johnson was published in two papers in *Genetics*, one in 1988 and another in 1990.[70] Another study, by Cynthia Kenyon at the University of California San Francisco, found that a different but similar gene, *daf-2*, also provided a one-shot means to extended lifespan in nematodes.[71]

From 1993–1995, Simon Melov and Gordon Lithgow, biologists at Buck Institute for Age Research, carried out stress experiments on nematodes with modified *age-1* and *daf-2* genes, demonstrating that they are more resistant to stresses capable of damaging DNA, such as high temperatures, ultraviolet light, and toxins.[72] Researchers Toren Finkel and Nikki Holbrook wrote a review of this and similar findings in *Nature*, in which they pointed out that every genetic mutation that accelerates aging also

---

67 Ibid., 44.
68 Ibid., 51.
69 Ibid., 71.
70 Ibid., 73.
71 Ibid., 74.
72 Ibid., 75.

increases vulnerability to these stresses.[73] The *age-1* and *daf-2* genes are part of a family of genes that controls insulin and insulin-like receptors. Changes to these have lengthened life in every species experimented on.[74] At Harvard University, researchers found an 18% increase in lifespan of mice lacking an insulin receptor in their fat cells, and these mice also ate 1 ½ times more than other mice but stayed thin and healthy.

Two other avenues of research into decelerating aging involve protection against free radicals and shielding mitochondrial DNA inside the cell nucleus. Free radicals are electrically charged molecules that react easily with other molecules inside a cell. They can strip electrons off strands of DNA, causing increasingly worse mutations over time. Fruit flies genetically modified for extra protection against free radicals live a third longer than others and are resistant to heat, starvation, and other stresses.[75] 99.95% of DNA is stored in the cell nucleus, except mitochondria—which is outside it with other cell machinery vulnerable to free radicals. Mitochondria ultimately decay due to a barrage of free radicals.[76] The nuclear membrane protects genetic information by means of repair enzymes that check and correct DNA. Aubrey de Grey, a gerontologist at Cambridge University who chaired a 2003 meeting of the Institute of Applied Biogerontology, has suggested moving the 13 genes responsible for mitochondria that are now outside the nuclear membrane into the cell nucleus so that they can benefit from its protective and corrective enzymes.[77] So far this has been done with one of them safely and without any apparent side effects on cell function.[78]

In 2003, Andrzej Bartke at Southern Illinois University successfully dilated the lifespan of his genetically modified laboratory mice to two-thirds longer than what had hitherto been considered the maximum

---

73   Ibid.

74   Ibid., 77.

75   Ibid., 82.

76   Ibid., 92.

77   Ibid.

78   Ibid., 92–93.

lifespan of the species.[79] Most lab mice do not live longer than two years. A few have lived into their third year. Bartke's mice died just short of their 5th birthday. Most importantly, *the mice lived out their extended old age without the diseases and fragility characteristic of it*. Researchers in the field of life extension refer to this as "compressed morbidity" — the fact that test subjects remain fit and active until their final days and then suddenly keel over without any apparent cause.[80] The molecular switches flipped in nematodes, yeast, flies, and mice exist in humans too, and the fact that it works in these species that are all so radically different from one another gives us good reason to think that it will work in humans as well.[81] Finally, granted that the minds and bodies of life-extended people also stay younger longer, there should be no serious danger of social stagnation and intergenerational conflict.[82]

In *Our Posthuman Future*, Francis Fukuyama argues that human biotechnology can and should be internationally regulated. One reason he takes this regulation to be possible is that international regulation has been accomplished in areas such as air traffic control, telecommunications, neuropharmacological drugs, replacement human body parts, nuclear technology, ballistic missile proliferation, bio-warfare research and human experimentation.[83] He argues that simply because there are lapses in enforcement of these international regulations does not mean that we should not have them at all.[84] The existence of laws is never a guarantee that criminals will not occasionally break them, but the occurrence of prosecutable offenses against the law make no case for abolishing law and order altogether.[85] Fukuyama thinks that the argument that any nation that would ban genetic engineering would only be crippling

---

79 Ibid., 72.

80 Ibid., 102.

81 Ibid., 78.

82 Ibid., 107.

83 Fukuyama, *Our Posthuman Future*, 188, 194.

84 Ibid., 189.

85 Ibid., 11, 189.

itself is baselessly fatalistic.[86] He does not think that regulation of human biotechnology requires transforming the United Nations into, or replacing it with, a planetary federal government whose resolutions are legally binding worldwide.[87] Fukuyama has in mind the gradual worldwide harmonization of policy among national institutions for the regulation of human biotechnology, of the kind that is currently underway amongst institutions regulating pharmaceutical products on a national basis.[88] This would have to be led by the world's most powerful and influential countries, especially those whose native natural rights traditions were eventually brought to bear on the formation of the international human rights regime of the United Nations and related institutions. For example, he believes that how the issue of human biotechnology is handled in US domestic law will go a long way in determining what kind of regulation there will be at an international level, if any.[89]

There are a number of problems with Fukuyama's argument for the possibility of effective voluntary international regulation of human biotechnology. In fact, in the course of his discussion of international regulation he unwittingly provides much of the ammunition against his own argument. First of all, international regulation on the proliferation of nuclear technology does not support his claim that human biotechnology can be effectively regulated. Nuclear weapons development requires a vast industrial infrastructure that is very expensive and difficult to hide as compared with what would be required for a technologically advanced nation to engage in a genetic engineering program.[90] Furthermore, although at the dawn of the nuclear age projections on the proliferation of nuclear weapons were very high compared with the number of countries that did develop them, and while the coercive international regulations of the NPT and the watchdog functions of the IAEA did discourage countries such as

---

86 Ibid., 11, 188.

87 Ibid., 194.

88 Ibid.

89 Ibid., 190.

90 Ibid.

Argentina and Brazil from going through with their nuclear ambitions, the very fact that extremely dangerous countries such as Pakistan and North Korea *were* able to develop nuclear weapons makes this example a case and point *against* Fukuyama's argument.[91]

Then there is the question of human experimentation. The case history here also actually undercuts Fukuyama's argument. The Nuremberg Code put in place after the Second World War was supposed to ban human experimentation without informed consent, but in fact, it was repeatedly violated, even in the United States, with the injection of live cancer cells into feeble chronically ill patients at the Jewish Chronic Disease Hospital, syphilis experimentation on the African-American Tuskegee airmen, and the Wilowbrook scandal in which mentally retarded children were infected with hepatitis. Eventually the Nuremberg Code was replaced by the 1964 Helsinki Declaration of the World Medical Association (whose membership is voluntary), which leaves regulation in the hands of scientists and doctors.[92] Especially given the precedent established by the latter, most scientists and doctors are likely to favor self-regulation — whose only real coercive force is the fear of scandals that could destroy reputations and mean the loss of funding.[93] However, the scientific elite is increasingly bound up with corporate research interests.[94] The genetic engineering industry of tomorrow is likely to wield the same kind of influence over doctors as the pharmaceutical industry does today in terms of 'self-regulation' of their Ritalin or Prozac prescriptions.[95]

Finally, the highly divergent attitudes of various countries towards biotechnology will pose considerable problems for the kind of non-authoritarian international regulation that Fukuyama has in mind. He does admit that international regulation of human biotechnologies would require the formidable task of forging of a consensus among very

---

91   Ibid., 189.
92   Ibid., 202.
93   Ibid., 215.
94   Ibid., 214.
95   Ibid., 209–210.

different cultures.[96] Countries in Continental Europe, Germany and France in particular, with their past experience of Nazism, are extremely conservative in what kinds of biotechnological research they allow.[97] On the other hand, perhaps due to Asian religious systems that lack a sharp distinction between humans and other animals, various Asian countries are quite open to human biotechnology.[98] As we shall see, China has a eugenics program in place that attempts to limit reproduction by low IQ persons, and it also allows harvesting of the organs of dead prisoners.[99] Singapore and South Korea have advanced biotechnology research infrastructures and have expressed interest in pursuing lines of research that may be banned in Europe and the United States.[100] Despite a dark history of horrifying human experimentation similar to the Germans, the Japanese also remain quite open to human biotechnology. Latin America and the Anglo-American countries (the UK, US, Canada, and Australia) fall somewhere in between Europe and Asia.[101]

Lapses in enforcement of laws against theft and murder, or even against bio-warfare research and non-consensual human experimentation, do not have a permanent and exponentially replicating effect. Fukuyama's case for the effectiveness of highly restrictive international regulation is vitiated on account of its being built almost entirely on analogies to significantly dissimilar cases. On Fukuyama's model of initially independent national-level regulation, followed by the gradual voluntary worldwide harmonization of national regimes, it is very likely that one or another country, probably somewhere in Asia, will get away with germ-line genetic engineering of humans for long enough to breed an entire generation of enhanced humans who will pass this pedigree down to their offspring in perpetuity. If a ban were to subsequently be put into place, it would only

---

96   Ibid., 12.
97   Ibid., 192.
98   Ibid.
99   Ibid.
100 Ibid., 191, 193.
101 Ibid., 192.

turn this nation's "new aristocracy" into the world's brightest and boldest in every major field of human endeavor. They would dominate the planet, culturally, scientifically, economically, and politically, which would likely also mean an overturning of the ban — unless they decided to retain their superiority over the rest of us through more traditional eugenic methods. Richard Lynn argues that this nation will be China.

A survey carried out from 1994 to 1996 found that 82% of Chinese physicians support eugenically motivated sterilizations, and up to 98% support counseling abortion in cases where the father has been diagnosed with a heritable disorder; by comparison only 5% of physicians and geneticists in Western democracies took this view.[102] Another survey asked geneticists and physicians whether they agreed that in addition to promoting the prospective mother's wellbeing, "An important goal of genetic counseling is to reduce the number of deleterious genes in the population." Whereas only 30% of respondents in NATO countries agreed, 100% of physicians and geneticists in China approved.[103] Another question asked whether eugenics is "the major goal of genetics," and again, *all geneticists and physicians in China agreed*. Meanwhile, only 13% of American physicians agreed that it was ethical to dissuade women from going through with their pregnancies if the fetuses in question were diagnosed with Down's syndrome.[104]

According to Richard Lynn the hostility to eugenics in Western democracies as compared to Asia is a consequence of the increasingly greater importance accorded to individual rights over social rights in Europe and North America. Lynn cites the political philosophy of John Rawls as an epitome of this development, quoting Rawls' magnum opus *A Theory of Justice* (1971) to the effect that: "Each person possesses an inviolability founded on justice that even the welfare of society as a whole cannot override. For this reason justice denies that the loss of freedom for some is made right by a greater good shared by others. It does not allow that the sacrifices imposed on a few are outweighed by the larger sum

---

102 Lynn, *Eugenics: A Reassessment*, 41.

103 Ibid., 294.

104 Ibid., 295.

of advantages enjoyed by the many."[105] Lynn rightly remarks that: "The acceptance of this premise would preclude the conscription of citizens into the armed services, the imprisonment of criminals, the detention of violent schizophrenics, the withdrawal of automobile licenses from habitual drunkards, and the like. It is a preposterous foundation on which to build a theory of justice and of the rights and obligations of citizens…"[106]

Still, it is on this basis that several European nations and US states have declared embryo selection illegal. The European Union has banned both cloning and genetic engineering of humans, and in 1990 a committee of the European Parliament determined that embryo selection "undermines our ability to accept the disabled."[107] Lynn predicts that for the medium-term future (several decades), development of more efficient contraceptives and the reversal of welfare policies, both of which only mitigate dysgenic effects, will be the closest thing to eugenic developments in Western democracies.[108] Positive eugenics will meet with the attitude of opposition epitomized by the 1989 declaration of the European Parliament that genetic techniques "must on no account be used for the scientifically and politically unacceptable purpose of 'positively improving' the population's gene pool" and that there should be "an absolute ban on all experiments designed to reorganize on an arbitrary basis the genetic makeup of humans."[109]

By contrast authoritarian Asian states are likely to implement the full range of biotechnologies that make a "new eugenics" possible. The best evidence for this is the fact that certain of those states have already adopted more controversial classical eugenics measures, such as immigration restriction, sterilization, and mandatory abortion. In Singapore in the 1970s Lee Kuan Yew, the country's prime minister from 1959 to 1990, instituted a state mandated classical Eugenics program consisting

---

105 John Rawls, *A Theory of Justice* (New York: Oxford University Press, 1999), 3.
106 Lynn, *Eugenics: A Reassessment*, 276.
107 Ibid., 264, 286.
108 Ibid., 290.
109 Ibid., 264.

mostly of positive measures such as tax rebates and state sponsored dating agencies.[110] Most significantly, the People's Republic of China initiated a classical Eugenics program in the 1980s, one far more draconian than any of those that Western nations attempted to implement in the early 20th century. This began in 1988 with the introduction of a compulsory sterilization and mandatory abortion program aimed at eliminating mental retardation in the Gansu province.[111] Then under the 1993 Eugenics and Health Protection Law, the Chinese government outlawed marriages for any persons afflicted with mental illness, venereal diseases, and hepatitis with the explicit aim of preventing them from propagating these diseases through procreation.[112] The Chinese Eugenic Law of 1994 established compulsory prenatal diagnosis for all pregnant women and demanded the abortion of any fetuses found to have genetic and congenital disorders.[113] It also authorized the nationwide compulsory sterilization of the mentally retarded and of those with other serious and readily identifiable heritable disorders.[114]

While the aforementioned policies are instances of negative eugenics, in 1999 the Chinese government also introduced policies of positive eugenics, such as the establishment of a state-run elite-semen bank, with donations obtained strictly from distinguished professors for the use of married women with infertile husbands.[115] The Chinese academic establishment may be at least as pro-Eugenics now, as elite American universities were early in the 20th century, when the respective presidents of Harvard and Yale, Charles Wilson and Irving Fisher, were both eugenicists and when Cornell University was doing Rockefeller funded Eugenics research.[116] Lynn speculates that Chinese adoption of Eugenics, which was

---

110 Ibid., 40–41, 294.

111 Ibid., 295.

112 Ibid.

113 Ibid., 41.

114 Ibid., 295.

115 Ibid.

116 Ibid., 27.

concurrent with the disintegration of the USSR and the Chinese decision to move towards a market system, is related to their recognition of the failure of communist ideology — which was, in principle, against Eugenics.[117]

As the biotechnologies of the New Eugenics are perfected and become relatively affordable for the increasingly wealthy Chinese state, there can be little doubt that they will be fully exploited. This development is likely to begin with *the replacement of procreation via sexual intercourse with mandatory pre-implantation embryonic selection*, which would be enforced by fitting 12-year old girls with a long-term contraceptive device to be removed only if their fitness to parent children, and that of their male partner, has been assessed and licensed.[118] *All embryos with genetic diseases and disorders would be screened out*, and only those on the high end of the general intelligence range possible for the approved parents would be selected.[119] Based on his assessment that this range consists of about 30 IQ points, Lynn projects an average increase in IQ of 15 points per generation.[120] Assuming he is right that East Asians already have the highest IQ of all populations (average 105), this would mean that *within several generations the Chinese would attain superhuman levels of general intelligence* as compared to Westerners, on average some 45 IQ points higher than people of European extraction. To demonstrate the consequences of this for scientific and technological progress leading to economic growth, Lynn cites the rapid post-colonial development of the East Asian economies of Japan, South Korea, Taiwan, Hong Kong, and Singapore in the second half of the 20th century. He attributes this to a mere 5 IQ point advantage over Westerners.[121]

Some eugenicists, and some critics of Eugenics, have argued against a precipitous universal rise in intelligence based on the danger that there

---

117 Ibid., 296.

118 Ibid., 299–300.

119 Ibid., 300.

120 Ibid.

121 Ibid., 305.

will be no one left to perform all manner of menial labor.[122] However, such an objection loses sight of two factors — first, the revolution in robotics and expert systems, and more ominously, the likelihood that China will attempt to establish the first truly global empire. As it stands, their rate of economic and military development has put the Chinese on the path to supplanting the United States as the world's superpower by the middle of the 21st century, of course in the guise of the 'international' *Shanghai Cooperation Organization* (which has already absorbed Russia and most of Central Asia). If this development were to be compounded by a "eugenics gap", it is quite possible that the Chinese will on the whole form the intellectual aristocracy of a global empire in which people of other nationalities — especially those in Africa, Latin America, and the Mideast — will fill jobs suited to those with lower intelligence and ability.[123]

Lynn also thinks that the Chinese will make use of cloning and genetic engineering. In his view, the highest strata of the Chinese global aristocracy might consist of clones of truly extraordinary scientific minds, as well as political leaders who ensure a much more orderly oligarchic succession than has hitherto been possible for authoritarian states.[124] Lynn speculates that China might only be a few years away from instituting a program for the cloning of elites, and unlike the Western democracies — where there would be a public outrage — the Chinese can bear the relatively few experimental failures necessary to perfect the process for use on humans.[125] Such a program would fit China's culture of ancestor worship. Lynn thinks that the Chinese will ultimately engage in research and development programs aimed at genetically engineering human beings to the point of their evolution into a new species capable of enduring the strains of long-distance space travel and colonization of other planets.[126]

---

122   Ibid., 289, 300.

123   Ibid., 308, 314–317.

124   Ibid., 303.

125   Ibid.

126   Ibid., 320.

As it stands, regulations of various human biotechnologies are completely inconsistent on a global scale.[127] Even within the European Union, there has not been agreement on whether to ban certain procedures, to legalize them without restriction, or allow them only under special circumstances. The most international agreement has been in the area of reproductive cloning, which is banned in 30 countries — but which has not yet been formally outlawed by the United Nations. By contrast, IVF is legal in most countries and is subsidized by Japan, Australia, and certain European governments. Genetic engineering has, to our knowledge, never been attempted and with the exception of a preemptive ban in a few countries, such as Germany, there is as yet little attempt to regulate it. The greatest diversity in how various cultures respond to the eugenic potential of human biotechnology has been seen in the area of PGD. In the United States it is legal and unregulated, with its proper use being left to the discretion of individual physicians who adhere to professional standards, such as the importance of follow up visits. It is also legal in China and India, except as a means to determine the gender of a child. In England, prospective parents must apply to a Human Fertilization and Embryology Authority whose task is to verify that they are in danger of passing a very specific disease down to their children. The attitudes of other European countries range from a total ban on PGD in Germany, to unrestricted use of it in Belgium — which has become a Mecca for those facing a ban at home. Two thirds of those who make use of PGD services in Belgium are from other countries.[128] Fukuyama is wrong to think that effective regulation of human biotechnologies with eugenic potential does not require establishing a world government.[129] On the contrary, if uses of the new eugenics are to be responsibly regulated, some kind of world policy-making institution with enforcement capability far more effective than the United Nations will indeed have to be established.

---

127 Naam, *More Than Human*, 144–145.

128 Ibid., 145–146.

129 Fukuyama, *Our Posthuman Future*, 194.

We now turn to the question of what criterion this world state should use in regulating the eugenic applications of various human biotechnologies. Fukuyama believes that the kinds of criticisms directed at the old *eugenics* are not relevant to the neo-eugenics of genetic engineering, for two main reasons: 1) the eugenics movement simply did not have the technological ability to effectively deliver what it promised; 2) the eugenics movement involved state coercion, whereas, certain Asian countries aside, most advanced liberal democracies — where the protection of individual rights is strong — are not likely to mandatorily impose genetic engineering on their populations.[130] The combination of technical efficacy and free choice left in the hands of prospective parents makes the kinder, gentler neo-eugenics immune to some of the most common arguments against the eugenics of the past. It also renders it more insidiously threatening. Fukuyama thinks that it is dangerous, in the name of *laissez faire* liberal individualism, to leave the choice to use neo eugenic technologies in the hands of parents.

Ramez Naam and Richard Lynn have both argued for the free market neo-eugenics policy that rightly concerns Fukuyama. Naam compares his laissez faire approach to human biotechnology to the market dynamics of new electronic goods. He admits that it is true that on the free market only the wealthy could afford the latest enhancements, but these would only feature minor improvements over older versions of a procedure at a much steeper cost for those who cannot wait for the prices to come down.[131] Although there is no inherent reason why Marijuana should not be as inexpensive as tobacco, having declared it illegal has made it hundreds of times more expensive per ounce. Likewise, Naam argues, bans on human biotechnologies would simply drive up the price on account of smuggling or other risks taken by service providers.[132] Furthermore, such bans are not likely to be enforced globally. Forcing people to travel around the world — as they already are — in order secure various genetic services

---

130 Ibid., 85–87.

131 Naam, *More Than Human*, 62.

132 Ibid., 62.

makes it a seller's market and drives prices up to the point that only the wealthy can afford enhancements for their children.[133]

Naam argues that embracing enhancement technologies is economically beneficial.[134] He claims that every 1% increase in productivity that may be gained by boosts in certain types of cognitive functioning would add $100 billion to the US economy, and every 1% reduction in health care costs on account of eliminating genetic diseases and disorders translates into $28 billion dollars to spend on something more productive. Naam claims that genetic enhancement would also be a means to address the impending Social Security crisis.[135] It is already clear that retirement ages are going to have to be raised and benefits cut. This would have a less deleterious impact if the elderly were in better health. Research into life extension has shown that the technologies involved actually compress morbidity as well, which Naam takes to be economically significant given that 47% of Medicare spending is allocated to the last 3 years of life. Extending youth by slowing aging and compressing the period of terminal decline would drastically cut health care costs worldwide.[136]

Lynn views embryonic selection as just another "consumer product" that "can be afforded by the affluent but not by the poor", such as the first cars and telephones, and he affirms the right of "citizens to spend their money as they choose" — just as some parents opt to pay their children's way through private education and others do not.[137] In the same vein, Lynn claims that if embryonic selection is legalized, there will still be some parents who *choose* entirely of their own will to procreate haphazardly. He thinks that these holdouts will be limited to only around 10% of the population, whereas I suspect that in very religious countries such as the United States, it is likely that even after legalization, at least 30–40% of the population would resist the 'diabolical' technological innovation.

---

133 Ibid., 146.

134 Ibid., 67.

135 Ibid., 102.

136 Ibid., 71.

137 Lynn, *Eugenics: A Reassessment*, 265, 286.

After a couple of generations, this opting-out by some will lead to a huge gap in intelligence, health, and ability between the "embryo-selected" and the "unplanned."[138]

If we combine this projection with the laissez faire free market attitude that Lynn shares with Ramez Naam, we realize that what is likely to emerge in an environment where neo-eugenic biotechnology is legalized but not subsidized or mandated is *the transformation of accidental economic class distinctions, which it is possible for enterprising individuals to transcend, into a caste system based on real genetic inequality*. This genetic caste society would be worse than a return to feudalism. Naam may be right that genetic enhancements are a potential solution to the impending Social Security crisis, but only if a far more serious situation of economic imbalance can be averted by implementing a subsidized social welfare program of compulsory neo-eugenics. Embryonic selection is not simply another "product", as Lynn takes it to be, because the embryo-selected children that it engenders are not "products." Introduction of the telephone and the automobile are not good analogies. A car or telephone is not enraged that it cannot compete with its peers no matter how hard it tries, because its parents could not afford to optimize it. It cannot demand the perpetual financial support of parents that could have afforded the optimization but decided against it.

What if parents' decisions to 'enhance' their children are informed by passing cultural fads, such as those that prefer a particular body shape (ultrathin girls) or a certain hair or eye color? While children are currently able to rebelliously resist the values imposed on them by their parents, say to change a name chosen under influence of a generational fad, genetic programming cannot be overwritten in this way. Fukuyama compares it to a tattoo that the children will have to hand down to all of their descendants, although I think he wants to say that the effects would be considerably more serious than even this trite example suggests.[139] If parents pay a great deal to have their child engineered for extraordinary

---

138 Ibid., 288.

139 Fukuyama, *Our Posthuman Future*, 94.

athletic ability and the child winds up wanting to pursue a scientific career and blames the parents for not endowing him with higher intelligence, this could frustrate both the parents and the child. If a child is designed for specialized ability in math or science and decides to become an artist instead, there will probably be much more resentment on the part of parents who paid for the increased IQ than there currently is when a young adult goes into an artsy field like painting or literature and scorns talents that his elders think could land him a lucrative job. Now one can say, 'well maybe he was really meant to be an artist', whereas then it will be clear that he was *literally meant* to be something else.[140]

The case against specialized germ-line engineering of humans is even stronger when it comes to the mixture of human genes with those of other animals to create human-animal hybrids or chimera. Fukuyama quotes biologists who think that producing an ape/human hybrid would be very scientifically rewarding, and he cites one experiment where human DNA was inserted into a cow egg and allowed to grow into a blastocyst before it was deliberately destroyed.[141] Producing such hybrids, even more so than cloning, would clearly violate any notion of human decency — whether it has some essential basis or is merely a linguistic-conceptual construct. It might even, as Fukuyama suggests, destroy any notion of "shared humanity".[142]

Ramez Naam shares this vision of a future where genetic engineering might actually lead to the artificial speciation of humanity. Witness the most dangerous and most revealing passage in his book *More Than Human*:

> At some point, one hundred years or one thousand years or one million years from now, our world and perhaps this corner of our universe will be populated by descendants that we might not recognize. Yet they will think, and love, and dream of better tomorrows, and strive to achieve them. They will have the traits most dear to us. They will be different in ways we cannot imagine. Most of all, these descendants of ours will be fantastically diverse. Where we are all alike in

---

140 Naam, *More Than Human*, 139.

141 Fukuyama, *Our Posthuman Future*, 206–207.

142 Ibid., 218.

the basic forms of our minds and bodies, they will exist in a plethora of forms. Humanity will have given birth, not to one new species, not to a dozen new species, but to thousands, or millions. We will have spawned a new explosion of life as sudden and momentous as the Cambrian explosion.[143]

Naam is talking about a speciation of the human race into "thousands, or millions" of "new species" with bodies *and minds* radically different from our own, and yet he assumes that they will *all* share exactly those traits that we praise most highly about humanity at present. First of all, it is clear that extant human cultures value very different traits in their respective populations. So who is the "us" in question here? Naam is really claiming that the values of the liberal and progressive elements of Western civilization — *his* values — will necessarily be preserved through an artificial speciation of homo sapiens into various other life forms in a cosmic analog of the Cambrian explosion. What kind of genetic modifications would transform humanity so radically that it would yield so many diverse post-human forms of life? Certainly, this would have to involve the insertion of the genes of other animals, or artificial genes designed on the basis of elements of the genome of animals very different from humans — perhaps including reptilian or avian or cephalopod elements. Moreover, Naam explicitly makes clear that these will not only be superficial changes of bodily form, but will lead to *minds* very different from our own — this would indeed have to be the case in order to justify calling these various post-humans instances of *new species*.

It is self-contradictory and frankly, absurd, to believe that somehow the core values of a particular human civilization — which may not even survive for our own species in the near future of *this* planet — will be preserved through all of these radical modifications in a plethora of divergent directions dispersed throughout colonized star systems. There is at least as much, and probably more, reason to think that other races branching off of a Humanity that has undergone artificial speciation would become as great a threat to the people of this planet as an advanced alien life form.

---

143 Ramez Naam, *More Than Human*, 198–199.

The only way to definitively avoid this nightmare scenario is to subject human biotechnology to regulations enforced by a world state. The legal mandate and enforcement capability of this single authority would not only have to extend over the whole of the Earth, it would also have to effectively govern (in advance) any potential future human colonization of extraterrestrial environments where genetic research and development might take place. This would have to take place within a time frame that obviates the danger that one or another nation, most probably China, will irresponsibly embark on its own state mandated and subsidized eugenics program. This nation might even lucratively offer specialized enhancements on the free market to prospective parents with the means to travel abroad.

We can get some sense of that time scale by looking at the rate of development of gene sequencing technology. Ramez Naam argues that the developmental rate of gene sequencing technology is tightly pegged to Moore's Law, since the two main sequencing techniques — the use of gene chips and linear DNA analysis — are both dependent on advances in computer technology.[144] Gene chips are surfaced with DNA strands that link up with the genetic material that they are exposed to, in such a way as to allow the base pairs of the latter to be read so that various genes can be identified. How many genes can be identified, and at what rate of speed, depends on how tightly packed onto the surface of the chip the DNA is, and refinements of this design capability are directly linked to developments in microchip manufacturing. In linear DNA analysis, enzymes are used to unfold a strand of DNA before it is sequenced with a laser base-pair reader whose data is processed by a computer. Improvements in processing speed markedly affect the quality of this type of sequencing.

The amount of DNA that can be sequenced at a given price is doubling every two years.[145] Whereas today we can identify only several dozen genes at a cost of hundreds of thousands of dollars, Naam projects that if the rate of progress of these sequencing technologies continues to follow Moore's

---

144 Naam, *More Than Human*, 122–126.

145 Ibid., 122.

law, by 2035 we will be able to sequence the entire human genome for just $1,000 and in only a day.[146] Six years later, in 2041 the price should drop to $100 and the sequencing will only take a few hours.[147] Leading companies like Affymetrix and US Genomics are projecting the growth of their own sequencing capabilities along these lines.[148] By the mid 21st century, at minimal cost prospective parents will be able to sit down in front of a monitor and flip through computer generated profiles of various embryos and click to choose their future child based on their facial features, hair and eye color, physique, IQ, certain personality traits, and the absence of every identifiable genetic disease.[149] This means that in order to effectively secure responsible regulation of the new eugenics, a world state needs to be established by 2050 at the very latest.

Not only would it have to enforce a single standard of human biotechnology regulation that bans genetic specialization, cloning, and any hybridization leading to artificial speciation, this world state would also have to pull off the feat of effectively enforcing compliance of its citizenry to subsidized mandatory non-specializing enhancements. This would meet with resistance both from religious conservatives who are willing to let their children be left behind, and others who want to buy specializing enhancements — to say nothing of a small but highly motivated and potentially wealthy anarchical elite of Transhumanists who want to push the boundaries ad infinitum.

This is all the more reason to consider acting while we can still persuasively shape the future, instead of waiting to violently react to it. Recall the words of a great leader of a very different kind than Chairman Mao, a visionary defender of liberty who once said:

> We stand today on the edge of a New Frontier. Beyond that frontier are uncharted areas of science… unconquered problems of ignorance and prejudice… We stand at this frontier at a turning-point of history. We must prove all over

---

146 Ibid., 122, 125.

147 Ibid.

148 Ibid., 124–125.

149 Ibid., 126.

again... Have we the nerve and the will? Can we carry through in an age where we will witness not only new breakthroughs in weapons of destruction, but also a race for mastery of the sky and the rain, the ocean and the tides, the far side of space, *and the inside of men's minds*? That is the question of the New Frontier.

That is the choice our nation must make — a choice that lies not merely between two men or two parties, but between the public interest and private comfort, between national greatness and national decline, between the fresh air of progress and the stale, dank atmosphere of "normalcy," between dedication and mediocrity.

All mankind waits upon our decision. A whole world looks to see what we shall do. And we cannot fail that trust. And we cannot fail to try.[150]

---

150 John Fitzgerald Kennedy, 1960 *Democratic National Convention* acceptance speech.

CHAPTER 4

Robotics and Virtual Reality

Biomechanics is by no means the only type of technology forcing the question of global regulation upon us. Robotics is another. Moreover many of these advances in technology are convergent and mutually reinforcing. For example, the most interesting development in robotics is probably biomimetic design. Instead of assuming that a robot can only be modeled on a human being or be some modification of an artifact or gadget conventionally used by humans, designers are now looking to a plethora of models from zoology to engineer robots that incorporate features of various animals that "jump, run, crawl, [and] do [other] things that nature does well."[1] Instead of going straight to an emulation of humanoid morphology, robot designers are now thinking in terms of eagle's eyes (for better surveillance), beasts of burden (as supply carriers), centaurs with four legs but a humanoid body (to solve the balance problem), and creeping or flying insects that could function as miniature recon drones entering homes through all of the openings that real insects do.[2]

As early as the 1970s the CIA had developed a bio-inspired drone about the size and shape of a dragonfly.[3] While they claimed that this dragonfly spy tended to get lost in grass, the research program that yielded this micro-robot continues to be funded to this day. Imagine

---

1 P.W. Singer, *Wired for War: The Robotics Revolution and Conflict in the 21st Century* (New York: Penguin Books, 2010), 91.
2 Ibid., 93.
3 Ibid., 117.

what it has achieved. The significant point, though, is that such classified technologies almost inevitably leak into the public sphere, even if the lag time is decades. So sometime soon corporations, and even some private individuals, are going to be able to spy on whomever they want with robotic insects. These critters may be loaded with electromagnets that allow them to recharge off your electrical outlets or light bulbs. More unsettling is the fact that some of these robots, modeled on insects with stingers, can be fitted with micro-syringes filled with poison.[4] A society in which such robots are widely available will have to be a maximal trust society. A legal system that punishes offenses after the fact would be ineffective. You would have to be comfortable with the idea of anyone being able to observe you doing anything at anytime. It would also have to be the case that no member of society wished harm or death upon any of his fellows, at least not to the extent of being willing to use deadly insect drones to assassinate that person in a relatively untraceable fashion. Otherwise, violent anarchy and mass panic would ensue.

Before taking a closer look at these socio-political implications, let us take a step back for a moment to consider the fundamentals of robotics. Contemporary researchers and developers use three key components to define a robot, all of which are irrespective of the level of sophistication of the robot in question. In order to be a "robot" a man-made device has to have: 1) *sensors* that scope out its environment and monitor changes therein; 2) *processors* of this information that, given its programming directives, decide how to respond to these changes (i.e. some, even minimal, level of artificial 'intelligence'); 3) *effectors* that are geared to act upon the world with a view to certain objectives by effecting a change in the robots environs. When all three components act together, we have a kind of artificial organism whether that is as simple as a robotic vacuum cleaner or as complex as a life-like android.[5] No matter how complex or artificially 'intelligent' a computer is, if it does not have effectors it is not a robot. Note that the effectors need not be humanoid. A driverless car is certainly

---

4   Ibid., 118.

5   Ibid., 67.

a robot, with its entire mobile frame capable of effecting various changes in the environment mapped by its sensors and processed by its computer system. On the other hand, robots do not need to be mobile. Factory floor devices with effectors designed for manufacturing labor on an assembly line certainly qualify as robots.[6]

Unlike in the categorization of human senses into sight, sound, smell, and touch, robotic sensors can be broadly divided into two categories: passive sensors and active sensors.[7] Passive sensors simply receive information *from* the surrounding environment, for example infrared sensors that collect surrounding heat source information. By contrast, active sensors send something out into the environment in order to aid in collecting information from it. One example of this is Laser Detection and Ranging (LADAR), which sends out laser beams and radio waves that map obstacles around a robot as they bounce back. In this process "perception" consists of a computer program parsing the data to identify and categorize the 'obstacles' in terms of their size and shape so that the robot may respond appropriately to various things it encounters. It should report rectangles of such a size and shape back to the base because they are "tanks", whereas large shrubs should be ignored (unless they are concealing insurgents).[8] A robotic truck deployed by a defense contractor in an occupied territory should stop if a child is in its path, but it should speed up and run over an insurgent who steps into the road with a rocket-propelled grenade launcher.

Wheels and tracks are still more common than legs as effectors that allow movement. Robotics engineers have had a very hard time with legs that maintain balance.[9] One effector for the combat models developed by iRobot is Metal Storm, a machine gun that is propelled by electricity rather than gunpowder so that it can fire ammunition so quickly and, on account of the robot's targeting, so precisely that it is able to deconstruct

---

6   Ibid.

7   Ibid., 80.

8   Ibid., 80.

9   Ibid., 32.

a target by shredding it. It can also put up a wall of bullets so quickly and so in synch with each other that the machine gun effector allows the robot using it to stop an incoming missile. Unfortunately for protestors, this weapon is also great for crowd control. One robot fitted with it, and without the drawback of human emotions, would be more effective than an entire unit of national guardsman or riot police.[10] It is worth emphasizing, in this regard, that once a robotic force of this kind is at the disposal of a sovereign government, the subjects of that government will be nearly powerless to stage revolts or insurrections. (This will become even clearer when we look at the variety of comprehensive surveillance technologies that are also emerging.) Consequently, within a few decades at most, we will need to have arrived at a form of government that we can basically accept as unimpeachable for the foreseeable future.

Real-world *Transformers* are also on the way. Scientists in Palo Alto and at MIT (funded by DARPA's "Programmable Matter" project) have designed robots that can fold and unfold like very complex origami in such a way as affords them the ability to shape-shift, appearing now as a snake and then as a spider.[11] The "claytronic" robots at Carnegie Mellon, currently in prototype stage, are pocket-sized *Transformers* that use electromagnetic forces to move, communicate, and share power when they are apart from each other but also have the capability of attaching to one another in order to form one big robot.[12] Contemplate the security implications of this technology. All of a sudden the furniture in a room could transform into a cadre of robotic assassins. Seemingly innocuous and apparently distinct objects carried onto an airplane could reassemble into a deadly weapon that would be ideal for hijacking. Of course, the same would be true anywhere there is a weapons check. Again, the only answer is not to have any terrorists.

---

10   Ibid., 83.

11   Ibid., 93.

12   Ibid.

That is ironic, given the fact that the Global War on Terrorism has been a major impetus for robotics R&D.[13] There are currently thousands of robots of diverse types deployed by the United States on various battlefields, and it is expected that before long conflicts will involve the deployment of tens of thousands of them.[14] 9/11 and the attendant establishment of the Department of Homeland Security has dramatically boosted investment in robotics research and development, at just the time that, on a theoretical level, engineering breakthroughs became a possibility.[15] The use of robots in fourth-world countries like Afghanistan, or the tribal areas of Pakistan, is particularly striking, since pre-modern battle tactics are still employed in these regions as well. You may have Predator drones targeting sword-wielding enemy combatants on horseback.[16]

During the recent Iraq war a group of Iraqi soldiers actually surrendered to a drone.[17] At the outset of the Iraq war there were no robotic systems involved in missions on the ground. Within a year of the invasion there were 150. Two years later, by the end of 2005, the number had risen to 2,400. This exponential increase had brought the number of robots deployed on the Iraqi battlefield to 5,000 by 2006 and 12,000 by the year's end of 2008.[18] These 12,000 robots represented 22 different types of systems.

The most famous (or infamous) of these is probably the Predator drone.[19] This Unmanned Aerial Vehicle is 27-feet long and lacking a cockpit. The nerve center of this flying robot is the proverbial rotating "Ball" located under its nose. This round device features a synthetic-aperture radar for penetrating clouds, smoke, or dust and it is also packed with two variable-aperture TV cameras, a daytime one and an infrared camera

---

13   Ibid., 107.

14   Ibid., 37.

15   Ibid., 39.

16   Ibid., 35.

17   Ibid., 57.

18   Ibid., 32.

19   Ibid., 32–33.

for seeing in the night. The Ball sends back recon data so precise as to be able to read a license plate from a height of two miles. Once a target is identified by means of its cameras or radar, the Ball's laser designator locks on it. Its composite material construction limits its weight to 1,130 pounds. It can fly up to a height of 26,000 feet for 24 hours. The pilotless plane is extremely cost effective; for the price of one F-22 fighter jet the Air Force can buy eighty-five Predator drones. The drones fly out of bases in the warzone, where ground crews of mechanics maintain them, but they are remotely piloted by people thousands of miles away back in the United States.

While Predators were initially designed solely for surveillance missions, their wings were eventually fitted with laser-guided hellfire missiles that can be activated by the drone's remote operator once a target has been acquired.[20] Nevertheless, some drones are still used mainly for surveillance purposes, in order to avoid incidents like the U-2 pilot Gary Powers being downed over the USSR and also to push the drone's capabilities beyond what a human pilot would be able to endure. Global Hawk is a 40 foot long surveillance drone that can remain airborne for up to 35 hours (much longer than a human pilot can endure) at an altitude of 65,000 feet (to put that in perspective, the maximum height of most commercial airliners is 35,000 feet).[21] On the other end of the spectrum in terms of size, are the Raven and Wasp drones deployed in Iraq by the United States military, which carry peanut-sized cameras and are, respectively, the size of a bird and an insect.[22] They are released by soldiers in the midst of combat zones to see what is over the next hill, around the next apartment block, or even to spy on someone through their window.

The term *robot* comes from the Czech language wherein it refers to a serf laborer bound to perform drudgery.[23] *Rabota* was an old Slavic word for a "slave." It was first introduced to the world through the 1921 play,

---

20 Ibid., 34.

21 Ibid., 36.

22 Ibid., 37.

23 Ibid., 66.

Rossum's Universal Robots. Written by Karel Capek and first performed in his native Czechoslovakia, the play about factory manufactured humanoid laborers (who eventually revolt against their human makers) was performed in New York City only a few years later where it introduced the word "robot" into the English language (in place of "mechanical man" or "automaton").

Ridley Scott's film *Blade Runner* addresses this theme at length, and although the replicants are "robots" their leader, Roy, at one point explicitly states: "We're not computers." The first thing to note about *Blade Runner* rolls by quickly in the text of its opening prologue: the Nexus 6 replicants that Deckard hunts down are referred to as cutting edge in "Robot" evolution. This is interesting because the commonplace conception of Robots is still of mechanical beings, whereas it becomes clear throughout the course of the film that these entities are the product of genetic design. They are biomechanically engineered artificial life forms. If there are aspects of human — and even animal — cognition that cannot be mechanistically modeled and replicated in a robot with a conventional neural network, it may be that the fusion of human consciousness with a drone robotic system by means of virtual reality may be the only route to strong Artificial Intelligence. The A.I. could be considered Posthuman insofar as integration into the drone system by means of cyberspace could profoundly alter the cognitive functioning of what was once a human being, especially if the user were plugged in during his developmental stage as an organism or even genetically engineered in a way that optimized his brain for integration with the drone system.

In a synergy of robotics and virtual reality, experimental systems are now being developed that allow a drone operator to experience piloting multiple drones simultaneously in virtual spaces.[24] The interface employs the hand gestural system featured in Steven Spielberg's *Minority Report*. Colonel Bruce Sturk, who runs a high tech lab at Langley Air Force Base saw the film and contracted Raytheon to develop the system for him. Raytheon in turn hired Spielberg's technology consultant in order to re-

---

24 Ibid., 69.

verse engineer the movie concept into a real world device. The "G-Speak Gestural Technology System" allows an operator to manipulate a computer interface by hand gestures alone, interacting with a projection of graphics and a free-floating keyboard rather than actual hardware. Sensors inside the gloves and cameras fixed around the operator's station track his movements, which are translated into computer commands. The Pentagon is even pursuing research into a total fusion of operator and drone by means of a haptic suit that translates stresses or malfunctions in parts of the plane into the sensations affecting correlated parts of the remote pilot's body.[25] This system would actually give the drone pilot such an immersive virtual experience of the drone's mission that he would begin to develop a different sense of his body and, in a way, it also means endowing the drone with a kind of sentience. The pilot who now has proprioception of the drone takes care of the device as if it is his own body, albeit a non-humanoid body.

Brain waves are generated by our neurons as they communicate with one another. These electrical signals could potentially be translated into digital code that a machine understands.[26] One important step in that process is to amplify the signal to noise ratio in order to isolate electrical activity representative of thought forms. These electrical signals leak through our skulls and can be picked up by an electroencephalograph or EEG. While this technology has already allowed people to move a computer cursor with their minds, its topical interface allows for a lot of interference from the outside. In his 1984 novel *Neuromancer*, William Gibson depicted hackers of the future plugging wires directly into their brains in order to "jack in" to a virtual world for which he coined the term "cyberspace."[27]

The first application of this technology was in the case of Matthew Nagle from Massachusetts, who in 2001 was paralyzed from the waist down. The BrainGate system, which uses a chip implanted in the brain

---

25 Ibid., 70.

26 Ibid., 71.

27 Ibid.

to isolate the signals that are on their way to arms and legs, eventually allowed Nagle to manipulate a robotic hand in order to do all kinds of things just by imagining doing them.[28] He can surf the web and change the channels on his TV just by thinking about it. Notice that if he was able to do this, the technology is already there for anyone to be able to directly connect her brain to the Internet and navigate it by thought alone.[29] Google searching will feel like asking a question within your own mind, except that an answer will pop up right away. But if our brains can be connected to machines in this manner, it also means that they can be connected to each other (via the mediation of those machines). If thoughts can be transformed, processed, and stored in the form of computer code, then they can also be hacked by others — potentially through the Internet to which our brains are connected. Technologically-mediated telepathy could, of course, provide soldiers with a tremendous advantage on a battlefield; they would begin to operate like a hive mind.[30] If you think this is far-fetched, consider that the U.S. government's National Science Foundation has taken the position that such technology is less than two decades away.

DARPA's Brain-Interface Project is one of its most well funded bioengineering research programs. Its aim is to develop a system that allows a pilot to control a plane with his thoughts alone so as to dramatically improve reaction time in combat.[31] While this interface could certainly be used with pilots inside actual cockpits, it could also be combined with drone technology in order to deploy remotely controlled planes that are thought-controlled via virtual reality. In fact, the latter is a more likely scenario since the extremely rapid movements of thought translated into airframes made of meta-materials that can withstand extreme stresses would probably yield g-forces that could crush a real pilot. It may even be possible to translate the thermal and infrared vision of the drone into a

---

28 Ibid., 72.

29 Ibid., 73.

30 Ibid., 74.

31 Ibid., 72–73.

graphic interface comprehensible by the pilot, one that endows him with a superhuman range of vision.[32]

Experiments that have planted chips inside people's body parts to convey sensor data from roaming robots suggest that such a degree of pilot-drone integration will lead to a new sense of embodiment. A common experience relevant to this is when people go from being novice drivers trying to park a car to being adept drivers whose bodily proprioception somehow extends to encompass the entire car so that they can almost feel the tightness of the parking space they are maneuvering into — which allows them to squeeze into such spaces with little or no thought, as if they were orienting their own body rather than a piece of machinery.[33]

Virtual Reality is just as much of a dual use technology as Robotics and various forms of biotechnology. There are all kinds of potentially positive applications of VR. Virtual reality is already used to test architectural designs, and as it becomes more sophisticated it is likely to be used to test all kinds of product designs, "hands on", before they are approved for physical manufacturing.[34] Consumers will also be able to do more than see a product from every conceivable angle, since haptic devices will allow them to get a feel for the product, and perhaps even smell it, before they decide to make a purchase.[35] Crime scene reconstruction in court rooms is another compelling use for virtual reality. Jury members at a trial will be able to be at the scene of a crime, examining whatever allegedly took place from the viewpoint of any supposed witnesses.[36] Virtual reality is already used on a small scale for surgical training, and the more accurate the rendering capabilities become the more it will be preferable to working with cadavers that do not give medical students as accurate a sense of what it is like to work with living organs. Such simulations could accurately measure, render, and feedback tissue damage to the virtual patient

---

32 Ibid., 73.

33 Ibid., 73.

34 Jim Blascovich and Jeremy Bailenson, *Infinite Reality: The Hidden Blueprint of Our Virtual Lives* (New York: William Morrow, 2011), 197.

35 Ibid., 199.

36 Ibid., 204.

undergoing whatever procedure the trainee is learning. In the near future, virtual reality could be employed by emergency room surgeons wearing light-weight head mounted displays to get inside the body of patients on whom they are performing laparoscopic surgery, or even to afford a particular doctor requested for his expertise to perform such surgery from a remote location while haptic devices make him feel like he has his hands on, or in, the patient.[37] Enter the VR ER.

However, in some ways, the potential threat to the human form of life from Virtual Reality is both more amorphous and more profound than that posed by any other emergent technology. It could become the most addictive drug in history. The enveloping of the 'real world' into the spider-web of Cyberspace could also utterly destroy privacy and personal identity, and promote a social degenerative sense of derealization.

When viewed against the backdrop of human life in simper times and more rudimentary cultures, advancing technology appears to perpetually inculcate new needs in those captured by its net.[38] A great many people who only a couple of decades ago were perfectly fine being incommunicado when not at home next to their grounded phone line, now feel extremely anxious without their cell phones. Connectedness to social media on the Web has become as vital a need for some as food, clothing, electricity, heat, running water and so forth. While Web-based social networks are much wider than pre-Internet social circles, it may also be the case that people spend so much time in virtual space that they neglect becoming someone whose persona is even worth "sharing."[39] Cutting someone's ties to the Web can leave them bereft and despondent. Sometimes being cut off from the Internet can even lead to violent outbursts. In October of 2009, a Swedish teenage boy stabbed a girl in his environs when an Internet service glitch interrupted his playing *StarCraft*.[40]

---

37 Ibid., 205.

38 Ibid., 171–172.

39 Ibid., 171.

40 Ibid., 181.

Of the two billion people connected to the Internet today, more than one billion are registered members of social networking sites.[41] An increasing number of these people spend hours online without a break, often instead of working or studying, and prefer to spend time on a computer rather than with friends or family; even when not on-line, they think about reconnecting.[42] They often find themselves on-line for much longer than intended, and sometimes try to conceal the extent of this non-essential Internet use from family, friends, and employers.[43] They use the Internet as a way to escape their problems, only to find, in some cases that their real relationships suffer more as a result. These are all signs of addiction to virtual reality.

People are used to thinking of addiction as always involving the abuse of some substance — like alcohol, nicotine, cocaine, or other drugs. However, the fact that some people can consume these substances *without* becoming addicted to them suggests that addiction has more to do with a personality type or a certain psychophysical constitution than it does with any given substance.[44] Being hooked on video games, mediated sex, social networking, and online gambling are various extant forms of virtual reality addiction.[45] Moreover, whereas one can stay out of bars, brothels, casinos, and crack houses, the paraphernalia for this type of addiction is inevitably in one's own home or even carried around on one's own person and is a dual-use technology that, at this point, is also necessary for being a productive worker.[46]

A Stanford University Medical School psychiatric study found that online addiction effects about 12% of the U.S. population.[47] As virtual reality becomes fully immersive and consequently much more compel-

---

41 Ibid., 175.
42 Ibid., 177.
43 Ibid., 179.
44 Ibid., 178.
45 Ibid.
46 Ibid., 183.
47 Ibid., 178–179.

ling, it is expected that the percentage of the population addicted to it will also increase significantly.[48] On-line gamblers will be able to enter virtual casinos as compelling as those of Las Vegas without leaving their homes.[49] How much more addictive will online sex become when people can buy a fully immersive, even tactile, encounter with avatars or agents that look like supermodels?[50]

The next generation of 'head-mounted displays' really will not be head-mounted because they will either consist of a lightweight pair of glasses equipped with a miniaturized projector that uses the full-spectrum of colored light to paint the VR rendering directly onto the user's retina or contact-lenses embedded with nanotechnology scale stereoscopic monitors and powered by body heat.[51] Imagine combining this with a nearly wire-frame nose-hugging smell-inducer, wireless ear-buds, a skin-tight haptic body-suit with additional attachable gloves and a headpiece that hugs everything except the front of the face, like in some diver's suits.

One can imagine virtual reality becoming so satisfying that its users withdraw from an increasingly disappointing 'real world.'[52] Video games were once played only by children (who grew bored with them pretty quickly) but are now popular amongst a much larger group of adults. That cannot only be on account of hyper-realistic graphics, but must have at least as much to do with the social networking dimension that allows people to escape into an alternate world populated by avatars, since the relatively crude Facebook game *FarmVille* now has nearly a hundred million users. Many of them are adults with children of their own. One woman admitted to not being able to sleep at night on account of worrying about her virtual crops; another shook her baby to death because its crying interrupted her attention to *FarmVille*.[53]

---

48 Ibid., 183.

49 Ibid., 186–187.

50 Ibid., 190.

51 Ibid., 58–59.

52 Ibid., 190.

53 Ibid., 180.

South Korea is a bell-weather country in terms of Internet addiction. It is such a serious problem there that boot camps have been set up to address the issue. South Korean youth are taken out of their homes and put into a group home, where they are forced to spend time together outside — doing work on lawns and pitching tents to camp out in.[54] Still, some spend their time outdoors daydreaming about getting back on-line. In 2005 one South Korean man collapsed of exhaustion and died after playing online in an Internet café for fifty hours straight.[55]

In addition to these issues of addiction, there is the threat to personal identity and the specter of an end to privacy. There are two types of entities populating virtual worlds, *avatars* and *agents*. The term *avatar* has been borrowed from Hindu spirituality in the context of which it refers to various, shape shifting, incarnations of a god or goddess visiting the earthly plane. The same deity can wear different bodies as it suits her.[56] Sometimes she can also become he, and vice versa — as in Shiva Ardhanari. So in virtual reality an avatar is any digital identity or on-line persona assumed by a user. By contrast, an *agent* may seem like a 'real person' but is actually an entity generated in virtual reality by the computer program. The most famous examples of these from popular culture are the "agent Smiths" of *The Matrix* trilogy. Depending on how good the program is at emulating human intelligence and socially responsible behavior, telling the difference between agents and avatars may be difficult. Just as a single person can don multiple avatars, multiple persons might also inhabit or construct a single avatar. A thirteen-year old girl committed suicide when she found out that people had concocted the boy she interacted with online with the aim of hurting her feelings.[57] The average age of avatar-donning gamers in networked virtual realities like *World of WarCraft* is now 26.[58]

---

54 Ibid.
55 Ibid., 181.
56 Ibid., 60.
57 Ibid., 5.
58 Ibid., 2.

Psychological experiments have been carried out involving body modification in a virtual mirror. A person wearing a head mounted display sees themselves in a virtual mirror as they redesign themselves or have their image designed for them.[59] There have also been experiments wherein users donning taller, more attractive avatars during on-line social interactions — for example in *Second Life* — have been monitored once back in the real world to see whether such experiences leave a lasting imprint on their behavior patterns.[60] In both cases, changes in virtual body image profoundly affected people's attitudes towards their real bodies and their interactions with others in the real world. Some people even wanted to have their biological bodies remade in the image of their avatars.[61] One can fluidly change the sex, age, weight, height and even the species of one's avatar.[62]

The possibilities of manipulating others through various misrepresentations of oneself are tremendous. A program might also be designed to manipulate people through agents that are so well crafted they are mistaken for avatars. Agents mistaken for avatars could be widely used to gather intelligence, a situation that would give the espionage term "agent" a whole new meaning.[63] There are already artificial intelligence programs that can so accurately record and replicate a person's speech-patterns, facial twitches, and idiosyncratic expressions such that very convincing agents can be programmed to pass themselves off as a digital doppelgänger capable of fooling those who know the person, at least for the duration of a short face-to-face conversation in virtual space.[64] When rendering becomes sufficiently high-definition with no perceivable lag, this virtual space could also be a real time video conference. The same technology could be used to record and store an accurate behavioral profile of one's girlfriend, so that after breaking up with her, a photorealistic avatar of

---

59 Ibid., 61.

60 106–107.

61 Ibid., 61.

62 Ibid., 63.

63 Ibid., 75.

64 Ibid., 149.

her complete with all of her expressive quirks can be reanimated as one's plaything in virtual reality.[65] A girl might have archives of everyone she has dated long enough for them to be modeled using A.I. programs that spy on Skype or FaceTime interactions to collect data for future reconstruction of her ex-boyfriends.

This calls to mind the reaction of certain primitive tribes to photo cameras; they were terrified that the demonic devices might steal their souls by photographing them. Maybe there was something to this intuitive response? As in the case of the robotic insect drones, the upshot here is that privacy and even personal identity cannot be protected by anything like the legal system that we have now. They would have to be implicitly and sincerely honored by everyone in the context of a maximal trust society. Moreover, given the unlimited geographical scope of these technologies, that maximal trust society would have to extend over the entire planet. Achieving such a society is not some ideal and distant, utopian goal. It is demanded by the need to maintain legal and political order in light of these technological advancements of the next several decades. So while it may at first seem more difficult to grasp how robotics or cybernetics and cyberspace require a world state, just as extraterrestrial energy acquisition and emerging biotechnologies do, in the end we can see how these technologies demand even more than a world state. They require, within 30 years, a world *society* or global village so cohesive that each member of it trusts the others more than almost anyone now trusts their spouse or best friend.

All of this becomes even clearer and more unsettling when one considers the increasing ubiquity of surveillance cameras in our technological society. In the near future, hidden cameras that are almost everywhere will be able to collect data enabling the construction, not just of photorealistic digital doppelgängers of oneself, but also ones that are behaviorally convincing.[66] The more ubiquitous the cameras, the more opportunity for tracking every quirk of one's behavioral pattern and range of facial and bodily expression for reconstruction and manipulation. Hidden micro-

---

65   Ibid., 151.

66   Ibid., 155.

phones could also collect audio and we already have the technology to perfectly mimic someone's voice and speech pattern. In a perverse inversion of signification, this type of data collection for the sake of programming simulations has been referred to as "reality mining" — as if 'reality' is there mainly to be mined as a raw material for the sake of virtual reality.[67]

The first military application of Virtual Reality was in flight simulators, and defense-related research and development of such devices was in fact one of the very first forms of virtual reality — long before its entry into the public domain.[68] Unlike the early flight simulators, which were for training purposes only, information networks and satellite communications now allow unmanned drones to be piloted by users who adopt the drone's viewpoint, and control its movements, by means of virtual reality — just as if it were a plane and they were in the cockpit. If robots who move like humans do become feasible, the same principle could be used to turn robotic soldiers on a real battlefield — possibly fighting a flesh and blood enemy army — into avatars of soldiers who are safe back at home.[69] The famous science fiction novel, and now film, *Ender's Game* involves a military force consisting of child soldiers who pilot drones and wage robotic wars by means of virtual reality.[70] James Cameron's film *Avatar* depicts the corporate-military colonization of another planet, predominately using avatars rather than subjecting flesh and blood persons to the dangers of environmental conditions in outer space.[71] Virtual reality is likely to be used extensively in space exploration and colonization. Short of developing cryogenic suspension technology, immersive simulations are likely to make long-distance space voyages, like the trip to Mars, bearable for the first generation of those who embark on such journeys.

Even members of an actual Lunar mining colony, for example, might spend a huge percentage of their time in virtual reality — not just to carry out their work by proxy through animating various robots at large on

---

67   Ibid., 164.
68   Ibid., 211–213.
69   Ibid., 213.
70   Ibid., 211.
71   Ibid.

the surface, but also as a much needed escape from the harsh and barren wasteland that is their workplace. Dutch professor Wijnand Ijsselstein has designed a virtual window system that can be installed in a room to make it appear just as if the room has a real window affording one a realistically shifting perspective view of any number of landscapes or cityscapes.[72] Imagine how compelling fully immersive virtual reality that engaged all the senses would be to people on the Moon. It would allow them to come back to the most beautiful on homely places on Earth at any time, and even to communicate with avatars of friends and relatives by such means. Liaisons and adventures with agents of various kinds are also imaginable, and possibly more practicable, since they would not require hi-fidelity instantaneous communication with real people back on Earth.

As we will see in the next chapter, mining colonies on the Moon are indeed a very real arena for the application of virtual reality—both as a psychological diversion from extreme isolation and also as an interface for the operation of drone robotic systems. All of the technological innovations that we have ben discussing thus far presuppose the continued development of industrial civilization. That, in turn, assumes the availability of an energy source at least as abundant and empowering as oil has been throughout the course of the past century. This should not at all be taken for granted in view of the imminent decline in petroleum past the global peak in oil production. I will explain why there is currently no alternative energy source or power-generating technology that can be safely substituted for petroleum in a way that takes up the slack to an extent that would ensure the survival of a growing industrial economy. The only solution will turn out to be the mining of Helium-3 from the lunar surface, as the fuel for clean-burning fusion power plants. So one impetus for research and development investment in virtual reality and robotic systems will probably be their use in the Persian Gulf of the 21st century.

---

72 Ibid., 225–226.

CHAPTER 5

Persian Gulf of the 21ˢᵗ Century

In the face of an impending planetary economic collapse past peak oil production, whoever controls the Moon's potential energy resources will also dominate the Earth. A return to the Moon for the sake of Helium-3 fueled fusion power can serve as a concrete conquest that challenges us to rethink fundamental concepts such as nationalism and international law. Off-world energy resource acquisition extends the political sphere beyond planetary dimensions in a way that threateningly encompasses the entire Earth. It is the perfect concrete technological advancement with which to begin radicalizing the realization that Carl Schmitt was reaching toward in his *Theory of the Partisan*. Namely, that revolutionary developments in technology that call human existence as such into question — on a planetary scale — might represent an exception to the rule that no cohesive and coherent political state can encompass the entire Earth and all of humanity. But in order to see just how necessary the latter is for the sake of regulating lunar resource exploitation for terrestrial energy, we must first confront the catastrophic socio-economic and industrial consequences of the impending energy crisis past peak oil production.

In the early 1950s a professor of geology at Columbia University by the name of M. King Hubbert, who had worked for the US Geological Survey, became the chief of research for Shell Oil. In this capacity, he developed mathematical models based on known reserves and rates of production and consumption.[1] Hubbert recognized that the rate of production would

---

1 James Howard Kunstler, *The Long Emergency: Surviving the End of Oil, Climate Change, and Other Converging Catastrophes of the Twenty-First Century* (New York:

mirror the rate of discovery — only decades later. Oil discovery in United States territory peaked in the 1930s and, he argued, that we should expect this to be mirrored by an arc of depletion past a certain point of peak production. He graphed what has now come to be known as "Hubbert's curve", with the point of peak production at the top of it. In the 1950s, Hubbert predicted that US domestic production would peak in 1970, and by 1971 data from oil companies proved him to be right: production had peaked in 1970 at 11.3 million barrels a day, and would fall by several percentage points a year thereafter, all the while demand increased.[2] While some tried to ignore the cold hard facts, they were driven home by the Oil Shock of 1973. The embargo orchestrated by the Arab states of the Persian Gulf in retaliation for Western support for Israel in the Yom Kippur war demonstrated that American domestic production was no longer sufficient to meet its own consumption demand and that the United States had become reliant on foreign imports, especially from Islamic countries.

Looking in the rearview mirror in the 1980s, Hubbert saw that world discovery had peaked in the 1960s and he expanded his model to project the date of global peak oil production. He initially estimated that it would occur between 1990 and 2000. According to Hubbert's model, if the rate of consumption were to remain at 27 billion barrels a year, then we will consume all of the two trillion barrels of oil that the Earth was discovered to have, within 37 years. However, a third to half of the oil left is likely to never be drawn out of the ground. The significance of "peak oil" is that past this point, recovery of remaining reserves begins to become economically unprofitable. What oil remains is harder to reach and of lower quality, so that the energy put in to retrieve it or to refine it begins to approach and then exceed the amount of energy to be gained from it.[3]

Although there have been projections of when we will have reached peak oil production, it is really only visible in a rear-view mirror. This is partly because major oil producers, especially those in OPEC, are

---

Atlantic Monthly Press, 2005), 41–42.

2 Ibid., 43.

3 Ibid., 49–50.

notorious for inflated assessments of their reserves. But more importantly, the peak would be marked by a period of price instability as increased cost initially causes an economic downturn and drives down demand. This period of instability — which appears to be happening *now* — would obscure the peak for a few years, until prices finally enter into a runaway upward trend reflecting the terminal downward slope of the production curve past peak.[4] Many experts in the field identified the period of 2000–2008 as that wherein world oil production would peak. These included: Colin J. Campbell, a geologist who worked for many of the giants of the oil industry, including Texaco, British Petroleum, Amoco, and Fina; David L. Goodstein, a professor of physics at the California Institute of Technology; Matthew R. Simmons, the CEO of Simmons & Company International, which is the chief investment banking firm serving the oil industry; Albert Bartlett, a professor emeritus of the physics department at the University of Colorado, Boulder; Jean Laherrère, a retired geologist for the French oil company, Total; Kenneth S. Deffeyes, professor emeritus of geology at Princeton University; Walter Youngquist, a retired professor of geology at the University of Oregon; L.F. Ivanhoe, coordinator of the M. King Hubbert Center for Petroleum Supply Studies in the Department of Petroleum Engineering at the Colorado School of Mines; Cutler J. Cleveland, director of the Center for Energy and Environmental Studies at Boston University; and, David Pimentel, professor emeritus of ecology and systematics at Cornell University.[5] Colin Campbell has pointed out that, although the petroleum industry does not take a public stance on peak oil because it would be bad for business (i.e. it would discourage investment in their enterprises), one can see their awareness of the decline in production reflected in the fact that they are no longer investing in new refineries, that they are merging, shedding staff and outsourcing people.[6]

Oil and its derivatives are vital to the modern way of life. Not only did oil as an energy source allow for the rapid development of the automobile

---

4   Ibid., 25.

5   Ibid., 25.

6   Ibid., 27.

by Henry Ford and others, its derivatives of asphalt and tar paved the way for a massive project of cross-country road building to service the new cars.[7] From 1914 to 1918, car ownership skyrocketed from 1.8 million to 9.2 million.[8] Meanwhile, oil powered plants allowed for electrification of growing urban areas connected to American suburbs, and to each other, by the highways.[9] The interstate highway system led to a dramatic disinvestment in U.S. cities and the relocation of mass populations to a suburbia that evolved out of the fast food joints, strip malls, chain stores, and office parks built beside the freeway off-ramps.[10] These stores are dependent on the shipment of goods over long distances by trailer trucks that demand cheap oil for the transportation costs to be affordable, and suburbanites are entirely reliant on inexpensive petroleum to commute to work.[11] The last great recoverable oil discoveries of the 20th century in Alaska and the North Sea drove prices down to unprecedented lows in the 1980s and 1990s.[12] This encouraged further suburbanization and sparked the gas-guzzling SUV craze.[13]

In addition to being refined into various grades of fuel, such as gasoline, diesel, kerosene, aviation fuel, and heating oil, for use in cars, trains, ships, planes, and homes, oil is also integral to the manufacture of numerous products that we have come to take for granted, including plastics, paints, pharmaceuticals, fabrics, and lubricants.[14] Our present agricultural practices are heavily dependent on petrochemicals, and these have affected soil to the point where even if we wanted to retreat to former agricultural practices the Earth would no longer yield what it once did. Furthermore, all of these applications of oil have greatly inflated

---

7   Ibid., 36.
8   Ibid., 41.
9   Ibid., 36.
10 Ibid., 41.
11 Ibid., 41.
12 Ibid., 29.
13 Ibid.
14 Ibid., 31.

the carrying capacity of the Earth, from fewer than one billion people to six times that number.[15] Nearly everything in the modern industrial economy sustaining this inflated population is either made or transported with some petroleum product.

In light of the North Sea and Alaskan reserves remaining to be exploited in that time, the oil embargo by OPEC in reaction to the Israeli role in the 1973 Yom Kippur war was a false alarm if one thinks of it in terms of the end of the age of oil. Nevertheless, conditions during that crisis offer us some insight into what the world after peak oil would look like in its early stages. By the time the embargo was called off in March of 1974 the price of oil had quadrupled, even though most of the 'embargoed' oil itself did continue to reach the United States, Europe, and Japan by more circuitous routes.[16] The United States was hardest hit, since it had historically been least dependent on foreign oil and had consequently not invested in electric rail systems, nuclear power, or other alternatives set in place by the Europeans and Japanese. In April of 1977, President Carter called the nation's energy crisis "the moral equivalent of war."[17] He called for the development of a whole host of alternative energy sources. However, careful studies of these alternatives have shown that they are either inadequate for meeting anywhere near the energy demands currently met by oil or they come with risks that only deepen and expand the scope of the energy crisis.

As it stands, natural gas accounts for approximately a third of the energy used in the United States.[18] The OPEC oil crisis of 1973 prompted many homeowners to switch from oil to propane or other types of natural gas to heat their houses during the winters.[19] However, there are some serious constraints on its potential expansion as a power source that might augment the loss of affordable oil. First of all, it should be noted that, like

---

15 Ibid.

16 Ibid., 46.

17 Ibid., 48.

18 Ibid., 102.

19 Ibid., 103.

oil, it gives off the major greenhouse gas carbon dioxide. So it is no better than oil in terms of its effect on the ecology of our planet, and it might be a great deal more dangerous to our urban and suburban landscapes. When mixed with air in concentrations of 5–15%, natural gas is explosive.[20] In a natural gas shortage, when pressure in pipelines to a certain urban or suburban area falls too low and pilot lights go out, there is a great danger that once the gas flow is restored there could be explosions that engulf homes and even whole neighborhoods.[21]

This is an especially pressing concern considering that natural gas reserves, at least in North America, have joined oil in passing their peak production and entering an arc of reserve depletion.[22] Between 2001 and 2003, there was an 82% decline in the supply produced by the 167 giant gas wells in operation at that time.[23] After the unusually harsh winter, in March and April of 2003 U.S. gas suppliers resorted to rationing — they cut off manufactures and power plants first, so that homeowners could remain warm even if they were sitting in the dark rather than working at their inoperative factories.[24] Of the United States, Mexico, and Canada, only the latter still has enough natural gas to meet its domestic needs and also export to the United States — which despite no longer being self-sufficient in its reserves is in turn, under NAFTA, bound to export its own natural gas to Mexico. Canada's ability to meet its own needs, let alone export to the United States, is steadily declining over time. As in the case of oil, the decline in production mirrors a decline in discovery.[25] Access to Canadian supplies could also be compromised by abrupt climate change. Furthermore, importing natural gas from outside of one's own continent is not like importing oil from overseas. Gas is pumped through pipelines at air temperature. In order to ship it across the ocean in tankers it first

---

20 Ibid., 102.
21 Ibid., 106.
22 Ibid., 104.
23 Ibid., 105.
24 Ibid., 106.
25 Ibid., 105.

needs to be turned into a highly pressurized super-cold liquid. Producing this liquefied natural gas (LNG) is very costly, and unloading it at specially designed port facilities is quite dangerous.[26] These port facilities, or for that matter the LNG tanker ships themselves, would make for very soft and yet spectacular targets in terrorist attacks.

The first source of energy to fuel modern industrial civilization was coal. Actually, the mining of coal had a reciprocal relationship with the rise of modern industry. The first coal-fired and steam powered pumps were developed to pump water out of flooded coal mines, and this laid the groundwork for the technology of the steam engines that powered ships, trains, and industrial machinery.[27] The downside of coal was, and remains, the terrible pollution of the ecosystem that it produces. The world's first major industrial center became infamous for its deadly "London Fog."[28] Coal burning, which still accounts for 25% of electricity production in the United States, is an even greater source of potentially climate changing greenhouse gas emission than oil.[29] While cars could not be run on coal power, we would not have to regress all the way back to coal-powered locomotives. Stationary turbines at coal power plants could run electric rail systems that replaced highways as the vital arteries of mass transit.[30]

Of course, this would still mean a total disintegration of the suburban way of life, and the return to more compact towns and cities that can easily be navigated by foot, bicycle, or electric subways. These would, in all likelihood, be very dirty cities with considerable health hazards. Coal puts out up to 20% of its original volume as solid waste. Coal burning has been linked to acid rain, asthma, mercury poisoning, and brain damage in newborn children.[31] Coal mining, especially the surface and strip mining that is now more common, ravages the natural landscape and

---

26 Ibid., 109.
27 Ibid., 116–117.
28 Ibid., 117.
29 Ibid., 117–118.
30 Ibid., 117.
31 Ibid., 118.

poisons groundwater. Cutting edge technology can be used to clean up coal emissions by filtering out heavy metals and particulates, but in an economy suffering from the shock past peak oil this is not likely to be affordable. Even if it were, such 'clean' coal technologies do not change the fact that coal plants will pump the atmosphere full of carbon dioxide.

What about the three very ecologically friendly power sources that are supposed to be already available alternatives to oil, namely hydroelectric, solar, and wind power? The basic problem with all three of them is the same: to use them on anything near a scale that might somewhat mitigate the loss of fossil fuels itself presupposes a fossil-fueled industrial background. The turbines, gears, and so forth of hydroelectric and wind power plants are manufactured with fossil fuels, as are the components of photovoltaic cells and the batteries to run them.[32] These components are vulnerable to the very forces of nature to which they are exposed to generate power, and once they wear down they demand replacement parts at regular intervals. The plastic, metal, lithium, and so forth wrought into the requisite components requires industrial processes uniquely powered by oil and natural gas. It might be theoretically possible to use hydroelectric, solar, and wind plants to power the industrial activity needed to produce and service the components of the plants themselves, but this would further divert energy from the useful output of these plants. It would push the energy invested over energy returned (EIoER) equation toward non-viability.

Given a drastically reduced fossil fuel base, these technologies could be used only on a very small and local scale. There are also considerable natural constraints on their applicability. For example, the United States currently receives 10% of its power from hydroelectric sources, and it has tapped everything but some remaining small rivers and creeks.[33] The maximum further expansion of this power source would be by 50% of its current yield, in other words it could grow to account for 15% of the current energy usage by Americans.[34] Installing solar panels where they

---

32 Ibid., 121, 123–124.

33 Ibid., 119.

34 Ibid., 120.

are vulnerable to frequent snow or rain, especially potential acid rain, is an exercise in futility on account of the rate at which the photovoltaic cells would corrode and need to be replaced by others produced through fossil fuel industry. Installing solar cells across vast desert terrain would be a great idea if those parts of the Southwestern United States were still viable for habitation, but once the end of cheap oil makes long distance suburban commuting and trucking supply of chain stores impossible, the population in these areas will be as sparse as it was before the industrial revolution. Wind farms are, similarly, viable only in certain areas and they produce nowhere near the amount of energy necessary to sustain heavy industry. Nuclear reactors could potentially power the heavy industry needed to manufacture the hydro, wind, and solar plant components and replacement parts, but as we shall see nuclear power has its own problems as a replacement for fossil fuel energy.

Before moving on to nuclear power, let me address a more high-tech clean energy source. Some have proposed switching from oil and gas to a hydrogen-fueled economy. Fuel cells, which have been used in manned spacecraft, employ a kind of reverse electrolysis to filter hydrogen through a catalytic metal membrane, where it combines with oxygen to produce an electric current that powers a vehicle; the only byproduct of the process is water vapor.[35] The fuel cells have no moving parts and are modular; their plates can be stacked to produce different amounts of power. Yet despite the elegance of their design and their ecological benefits, these fuel cells pose some serious problems of storage and transport.[36] On account of its low atomic weight, hydrogen takes up a lot of space, which means that it has to be compressed and stored in high-pressure tanks in order to be used in automobiles. This compression to something on the order of 10,000 pounds per square inch takes a lot of energy and it poses a considerable safety hazard. Even if the tank itself were to survive a car or truck crash, the more delicate pipes trailing out from the tank probably would not and hydrogen would escape at high pressure. The problem is

---

35   Ibid., 110.

36   Ibid., 113.

that mixtures of hydrogen and air combust with energy inputs of less than 10% those required to ignite gasoline exposed to the air. Hydrogen is also capable of self-igniting if it is highly pressurized, as it would be in these tanks, so that at impact it decompresses out of broken valves at a very high rate of speed (i.e. with a high kinetic energy). Hydrogen also leaks very easily. Due to its low atomic weight it can escape out of much smaller holes than gasoline, and it is extremely corrosive — so that if it does leak, it can rapidly disintegrate the pipes, valves, and seals of the vehicle in which a hydrogen fuel cell is installed.

In order to replace gas stations with hydrogen fuel cell charging stations — a retrofitting task that would probably take longer than a decade assuming we could even afford it — liquid hydrogen would have to be trucked around in large high-pressure tanks.[37] The same truck that can carry 25 tons of gasoline can only carry ½ ton of hydrogen. An average filling station on a freeway sells 25 tons of fuel a day and it can be serviced by one 40-ton gasoline truck. In order to deliver the same amount of energy to that station, to power the same number of cars that refuel at it, one would need more than 20 hydrogen trucks.[38] Currently, 1% of trucks are gasoline or diesel tankers, whereas switching over to fuel cells would mean that 17% of trucks would be hydrogen transports. That means roughly one out of six accidents involving trucks would be one where a hydrogen tanker, for the reasons explained above, would inevitably turn into a high-yield explosive device.[39] This would trigger firestorms spreading through tens of combustible cars on freeways. We would quickly find that the safety hazards associated with such a technology exceed those that made the early Zeppelins non-viable. It is possible that exotic materials, perhaps involving nanotechnology, could be used to fortify cars and trucks against such safety hazards associated with their hydrogen cells and tanks,[40] but in the near-term future this would mean that being a

---

37 Ibid., 113–114.

38 Ibid., 114.

39 Ibid.

40 Ibid., 115.

motorist becomes a privilege of the wealthy — in which case, why not just run the cars of the elite on the increasingly expensive gasoline that they alone will still be able to afford?

Another major problem is that hydrogen technology cannot be developed on a large scale for use as a carrier of energy without being fueled by some other energy source in the first place. As it stands, it takes more oil and natural gas to produce hydrogen fuel cells than the cells would generate.[41] Hydro, wind, and solar power are not nearly abundant enough sources of energy to meet the industrial production demands of hydrogen fuel cell technology. The only viable source of large scale power to get the switch over to hydrogen going is nuclear energy, so that "hydrogen economy" is largely a euphemism for a "nuclear economy" — a world where nuclear power has expanded to replace oil and gas, and is in turn used to power hydrogen fuel cells for cars, trucks, and other machines that cannot be fit with a nuclear reactor or directly connected to one through power lines. Speaking of power lines, or gas lines for that matter, hydrogen fuel cells would not offer an alternative to replacing oil and gas to heat buildings, so in this sense as well, it is a technology that would have to be developed and implemented in tandem with a vast expansion of the nuclear energy infrastructure.[42]

Of all the readily available energy alternatives to oil, nuclear power alone has the capacity to sustain our current level of industrial development and buy time for our technological civilization to arrive at some other breakthrough.[43] Currently the United States only produces around 20% of its energy from nuclear power, as compared to France where 70% of the energy sector is nuclear powered.[44] In fact, much of the rest of the French energy infrastructure is hydroelectric, with the industry for producing hydroelectric components being augmented by nuclear power. This is a model that we could adopt. Forcing an American populace with

---

41 Ibid., 111.

42 Ibid.

43 Ibid., 141.

44 Ibid., 140.

a tremendously complacent sense of entitlement out of suburbia and the gas-guzzling cars that they use to navigate it, and into compact urban areas serviced by electric rail would be a challenge, but it is one that pales in comparison to simply allowing the current system to collapse before having replaced it with an atomic alternative. Even if the production of components for nuclear power plants in some cases may require fossil fuel technology, coal or synthetic oil produced from it could be used on a small scale for this purpose. The energy thus invested would be greatly exceeded by the energy returned through the fission of widely available and naturally occurring uranium, which produces two million times more energy per unit of mass than oil does.[45] In other words, a pickle jar's worth of uranium supplies enough energy to an entire family for their whole lifetimes, and that jar's worth is relatively cheap; uranium costs about $15 a pound. Even poor countries such as North Korea can afford the enrichment and reprocessing facilities needed to refine it to the requisite purity for use in power plants. The basic centrifuge technology for uranium enrichment is now more than 60 years old.

Unfortunately, the radioactivity of uranium and other nuclear isotopes is deadly and the non-biodegradable waste produced by using them has a very long half-life. Not only are nuclear reactors themselves vulnerable, but the structures storing their spent fuel rods could also be compromised. While nuclear power plants themselves do not produce any atmosphere pollutants, such as greenhouse gasses, so long as it is operational a damaged nuclear reactor core or spent fuel storage facility could not only fill the atmosphere with radioactive fallout, it could also poison groundwater and spread by this means as well. That translates into a huge rise in cancer cases and birth defects, as well as destruction of arable land, livestock, and/or ocean biodiversity relevant to the fishing industry. So far we have been extremely fortunate. There have only been three major nuclear power plant catastrophes. Yet, each of them has demonstrated the tremendous cost of a switch over to nuclear reactors as our main source of energy.

---

45 Ibid., 141.

The first of these incidents was only a near-disaster, when in 1979 a meltdown was narrowly averted at the Three Mile Island plant in Pennsylvania. Although no deaths resulted from the radioactive gases that were vented, public comprehension of the scale of the potential catastrophe if the reactor had actually gone into meltdown killed a plan by the Carter Administration to wean us off increasingly unreliable Middle East oil by a large-scale expansion of the nuclear power industry (Carter himself was a nuclear engineer).[46] The Chernobyl meltdown in the Soviet Union on April 26, 1986 irradiated not only Russia but most of Eastern Europe as well, leading to thousands of cancer deaths and dramatic rises in birth defects that continue to this day.[47] Yet it may be that the worst crisis is the most recent. The Fukushima meltdown in Japan, which involved earthquake damage to spent fuel rod storage facilities in addition to two reactor cores, has been estimated to have exceeded Chernobyl in the amount of radioactive fallout that it dumped over East Asia and even all the way across the Pacific to the Western United States. The scale of the ecological catastrophe was hidden at first by the Japanese government, which has since been forced to admit that the cleanup might take 50 years. The environs of Fukushima, just north of Tokyo, have become an uninhabitable dead zone. In a country where seafood is the staple of the national diet, Japanese fisheries have been contaminated and we are likely to see another generation born with atomic mutations of the kind that appeared for decades in the environs of Hiroshima and Nagasaki. The total deaths in radioactively induced cancers, both in Japan and elsewhere in East Asia, might eventually climb into the hundreds of thousands.

The Chernobyl accident is supposed to have been caused by human error and Fukushima was the outcome of a natural catastrophe, but what has to be considered above all in terms of any plan to replace fossil fuels with an expansion of nuclear power is the potential for terrorist attacks on nuclear plants as well as the dual-use capability of nuclear energy. If we affirm nuclear power as the near-term future, how can we effectively

---

46 Ibid., 143.

47 Ibid.

deny it to other nations that will also have to contend with the end of oil, especially those Muslim countries with economies structured around oil. Nuclear power plants were supposedly the original targets of the 9/11 hijackers. These plants have no air defenses, and they are poorly guarded at the ground level as well. In a world where wars over the last few drops of recoverable oil rage across Muslim lands, or increasingly severe and unpredictable storms routinely wreak havoc on infrastructure, how viable is nuclear power in the United States or for that matter even in France?

The dimensions of the global energy crisis are even more serious than we have grasped them in the foregoing analysis. A very reasonable projection of growth in energy demand by 2050 is the point of departure for Apollo astronaut and geologist Harrison Schmitt's advocacy of a return to the Moon to mine it for Helium-3 fuel for use in nuclear fusion power plants. To arrive at this projection, Schmitt takes the expected increase of world population to around 10 or 12 billion by mid-century together with the aspirations of developing countries for standards of living comparable to those of developed nations. These two factors are linked insofar as there is a definite trend of population stabilization commensurate with increased energy consumption, which in turn has thus far been the only reliable measure of improved living conditions.[48]

The current average use of energy per person per year on Earth as measured in the equivalent of oil is 12 barrels, which translates into a global annual consumption of more than 70 billion barrels of oil equivalent (BBOE).[49] The difference in standard of living between say, the United States, and the global average is reflected in the fact that the per capita energy consumption of an American is a little over 60 BOE. Unless energy consumption increases to a certain threshold across the developing world, total energy demand will actually rise even more significantly as scaled to a global population that continues to increase exponentially rather than level off at 12 billion. Accounting for an elevated standard of living that would stabilize population worldwide by mid-century, with an average

---

48 Harrison Schmitt, *Return to the Moon: Exploration, Enterprise, and Energy in the Human Settlement of Space* (New York: Copernicus Books, 2010), 26.

49 Ibid., 25.

standard of living still far below that of Americans, we would have to project an eight-fold increase in annual energy production. Should ambitious world powers with huge and growing populations, such as China and India, be dead set on attaining a standard of living for their citizens comparable to living conditions in the United States, this conservative estimate would have to be scaled up to at least a 12-fold increase in available energy resources by 2050.[50]

Whether an eight-fold increase is overly conservative, in the event that China follows the development curve of a country like South Korea, is really immaterial. As we have seen, the fact is that even if the global population were to hypothetically freeze at current levels, the peak in fiscally retrievable fossil fuels — not just oil, but also natural gas and coal — would not support the present standard of living in the developed world up to, let alone beyond, the middle of the 21st century. Furthermore, as fossil fuels become scarcer their value as fossil chemicals used in agriculture and a wide variety of consumer products will also increase.[51]

Controlled fusion reactions offer great power generation promise if a fuel source could be secured that is more benign than tritium or other radioactive isotopes. Unlike fission power plants, fusion reactors could in principle be clean burning. Helium-3 — a light isotope of the birthday balloon gas helium-4 — is a clean fuel for nuclear fusion power.[52] The first fusion of helium-3 and deuterium was demonstrated as early as 1949, but its potential as an energy source was virtually ignored because a commercially viable supply of helium-3 is not found on Earth.[53] A very small amount of helium-3 is emitted from mid-oceanic vents and amounts just barely sufficient to do theoretical research can be purchased, at a very high cost, from the United States or Russian governments who obtain it from processing the tritium in decommissioned nuclear weapons.[54] It

---

50 Ibid., 27.
51 Ibid., 31.
52 Ibid., 44.
53 Ibid., 79.
54 Ibid., 80.

is, however, abundant on the Moon. This occurred to researchers at the Fusion Technology Institute of the University of Wisconsin-Madison in 1985.[55] They reasoned that over the course of nearly four billion years, the absence of an atmosphere would have allowed the solar wind to deposit helium-3 in the lunar regolith. Whereas fossil fuels are produced out of organic materials by relatively low pressure and temperature chemical reactions, helium-3 is produced by extremely high energy fusion reactions in the Sun; these fuse its hydrogen into helium and other elements that stream away from the Sun along embedded magnetic lines of force extending throughout the solar system, including Earth's moon.[56] The density of the solar wind that carries helium-3 to the Moon varies greatly, periodically spiking during solar flares or coronal mass ejections.[57] Since the Moon has no atmosphere the solar wind ions bombard mineral and glass particles at its surface, penetrating these individual grains to a depth of a few ten thousandths a millimeter.[58]

During Apollo 11 Neil Armstrong filled a rock box with 17 scoops of bulk regolith from the landing site in the Sea of Tranquility.[59] Based on an analysis of samples collected during the Apollo program, it has been estimated that helium-3 is present in concentrations of at least 20 parts per billion in undisturbed lunar soil.[60] While this may not seem like much, when compared to the energy equivalent value of coal, helium-3 could produce so much power per volume that it is valued at $40,000 an ounce. To put that in perspective, gold is now worth around $400 an ounce. So helium-3 is, per mass, about a hundred times more valuable than gold.

The start-up cost of establishing a mining and processing operation that is of a sufficient scale to financially break even after five years from the initiation of production may require an investment as small as $2.5

---

55   Ibid., 79.
56   Ibid., 78.
57   Ibid., 78–79.
58   Ibid., 79.
59   Ibid., 84.
60   Ibid., 44.

billion. This very conservatively assumes half the number of helium-3 fusion power plants coming on line each year as the number of fission nuclear power plants that became a new energy source in the United States between 1973 and 1990, which is a fraction of the rate at which new nuclear plants came on line in France during those years.[61]

Ultimately, the theoretically sound fusion of helium-3 with itself should be empirically demonstrated and designs for reactors that are based on this type of reaction, and that will yield no radioactive waste whatsoever, can be optimized for industrial production.[62] Although the elimination of waste disposal costs might favor pure helium-3 fusion, this would have to be weighed against the fact that terrestrially abundant deuterium is much more inexpensive than lunar helium-3 and the fusion of helium-3 with itself has a lower reaction rate than its fusion with deuterium.[63] Experimental helium-3 fusion research reactors have been constructed at the Joint European Torus (JET) and the University of Madison-Wisconsin, but their limit of generating around 16 MWe of power disqualifies them from being prototypes for a commercially viable plant. Research and development costs for construction of a commercial prototype helium-3 nuclear fusion plant are estimated at $6 billion.

A sum just slightly less than that would take us back to the Moon with the same payload capacity as the Saturn rockets used for the Apollo program. Development of a Saturn VI that substitutes 1960s components with cheaper and more elegant modernized ones while delivering 100-ton payloads to the Moon, is supposed to cost $5 billion. This takes into account sufficient funding for design R&D as to avert the lack of confidence crises that led to repeated post-disaster delays in the Space Shuttle program, launch delays that simply could not be tolerated in the case of privately-financed and investor-driven lunar mining operations.[64] Return to Earth costs would be minimized by any private enterprise recruiting

---

61 Ibid., 73.

62 Ibid., 45.

63 Ibid., 66.

64 Ibid., 129.

only lunar employees ready and willing to become settlers of growing Moon colonies initially centered on mining.[65]

Many of the mining operations can and should be automated using robotic and telerobotic systems.[66] For example, telerobotic operation of the miner units from indoor consoles and robotic automation of processing would probably be more cost effective, with risky exterior human activity limited to periodic maintenance and repair interventions.[67] Nevertheless, permanent settlement is essential to effective and sustained mining of helium-3 resources on the Moon.[68] In addition to detailed studies of lunar colonization done in the 1960s, such a project could draw from decades of relevant experience from Apollo, Skylab, Mir, Spacelab, the Space Shuttle, and the International Space Station as well as at polar research stations on Earth.[69] Unmanned spacecraft would frequently send helium-3 back to secure locations on the Earth in low mass shipments small enough to accommodate the risk of losing any given one, possibly in a cryogenic liquid form stored close to absolute zero but probably in a more low cost pressurized gas form.[70]

A number of byproducts from the helium-3 extraction process have commercial value and can help to finance the cost of mining and processing. These include hydrogen, water, and oxygen very valuable to customers in space, either in Earth's space stations or on a variety of vehicles venturing from the Moon's 1/6th launch-gravity onto Mars or deeper into the solar system.[71] Assuming that mining operations lead to a certain level of lunar colonization, and even tourism, there ought to be sufficient customers in space for these derivatives. Hydrogen and oxygen produced on the Moon could also be used for fuel cells to power mining operations there,

---

65   Ibid., 46.
66   Ibid., 110.
67   Ibid., 114.
68   Ibid., 109.
69   Ibid., 110.
70   Ibid., 128.
71   Ibid., 46–47.

in conjunction with solar power.⁷² Not taking these longer-term benefits into account, the total cost of launching massive payloads and employees to the Moon, setting up helium-3 mining and processing facilities there, and constructing the first commercially viable helium-3 nuclear fusion plants back on Earth comes to less than $15 billion.⁷³

This is a cost comparable to that of the 1970s TransAlaska Pipeline or the 1980s EuroTunnel. It is also competitive with the cost of finding and exploiting helium-3's energy equivalent in oil during the period leading up to global peak petroleum production. It is less than 1% of what the United States wound up spending on the invasion and occupation of Iraq from 2003–2011, a time frame comparable to that of this project if it were to be given the level of priority that John Kennedy secured for the Apollo program. Indeed, in order to meet the rising electricity demand in this eleventh hour before the catastrophic energy crisis past peak production of fossil fuels, research and development of helium-3 fusion cannot take the form of gradual development of current D–T fusion technology; it would have to proceed on the scale and pace of the Manhattan Project.⁷⁴

The most significant obstacle facing any venture to mine the Moon for helium-3 is not technological or economic in nature, it concerns lunar law and order. Harrison Schmitt acknowledges that if there is to be permanent settlement of the Moon, as the mining colony begins to self-replicate people will want legal protection for their families and, what at that point can only be called, their homesteads — especially since multiple nations and their cultures will have a lunar presence.⁷⁵ Indeed, Schmitt goes so far as to explicitly acknowledge the eventual propagation and cohesion of distinct religious communities on the Moon.⁷⁶ Past a certain point corporate contracts delineating the legal responsibilities of employees will be grossly inadequate. Schmitt looks to a number of extant international treaties and

---

72 Ibid., 125.
73 Ibid., 47, 142.
74 Ibid., 72.
75 Ibid., 313–315.
76 Ibid., 313–314.

the debates surrounding them as he argues for a permissive legal framework to govern lunar resource exploitation. The most important of these treaties concern Antarctica, international waters, and two attempts at a treaty that would extend international law to the Moon.

The Antarctic Treaty of 1959, which entered into force in 1961, effectively prohibits any territorial claims by countries or extensions of their national sovereignty to any part of Antarctica — pledging that "…Antarctica shall continue forever to be used exclusively for peaceful purposes and shall not become the scene or object of international discord."[77] It was signed by all of the major powers of the time and has since developed into a legal framework that encompasses all 38 nations that have anything to do with Antarctica. So attendant to this treaty, designations like "Queen Maud Land" (or even more so, New Swabia) that have ring of national territorial claims, are to be considered quant relics of a bygone era of colonial exploration and exploitation. Since a principal cause for the "international discord" so explicitly prohibited by the Antarctic Treaty has, throughout history, been resource exploitation, two subsequent conventions have addressed this question. In 1988, with the increasing awareness of the potentially rich subsurface resources of Antarctica and its continental shelf, an Antarctic Convention on Mineral Resources took a step towards regulation of mineral exploitation sponsored by ratifying parties.[78] Severe criticism by environmentalists followed and the mineral convention was quickly superseded by the Antarctic Protocol on Environmental Protection of 1991, which is essentially prohibitive of substantive resource exploration and extraction on the grounds of its environmental impact. Schmitt speculates that nations party to this protocol may pull out of it once resources of sufficient value become accessible in Antarctica, and he unequivocally rejects modeling a lunar legal framework on the international regulatory approach taken to "that frozen continent".[79]

---

77 Ibid., 279.

78 Ibid., 279.

79 Ibid., 280.

The ecological impact would certainly be dis-analogous. None of the changes brought about by mining lunar helium-3 would be noticeable from the Earth, even using the best telescopes.[80] The largest craters destroyed would be 20 meters in diameter, far too small to effect a noticeable change in the lunar features as observed from Earth. Any alteration of the Moon's solar reflectivity by even the most large-scale mining would also be so slight as to be effectively imperceptible. Dust and other debris ejected during operations would simply land again nearby, and in any case it would be less than the disturbance to the surface already caused on a regular basis by daily meteor impacts. The lack of wind and water on the Moon will also localize human liquid, solid, and gaseous waste, which will be recycled for greenhouse agriculture. Out of necessity damaged hardware will also be stored in an inventory and cannibalized for new purposes as the need arises, instead of being buried as lunar trash. Finally, it must be remembered that the clean fusion energy that these mining operations would bring online would have a significantly positive impact on the Earth's ecological conditions by rendering polluting forms of power production completely obsolete.

The Law of the Sea is another area of international law on the Earth relevant to setting a precedent for lunar resource development. The United Nations Law of the Sea Convention of 1982 was prompted by discoveries of large mineral and other deep seabed resources with substantial industrial value, and a need to internationally arbitrate access to these in cases where they lie beyond any given nation's 12-mile limit of territorial waters.[81] Oil in the rapidly de-glaciating Arctic Ocean would be one example. Due to what the United States found to be excessively burdensome demands on sharing any exploited resources with the international community, through high fees on extraction and mandated transfers of technology, President Reagan not only refused to ratify the convention but between 1980 and 1983 created a Deep Seabed Hard Minerals Act granting the US National Oceanic and Atmospheric Administration (NOAA) authority to

---

80   Ibid., 121.

81   Ibid., 289.

license US nationals for deep seabed mining and, by his executive order he established an "Exclusive Economic Zone" that effectively, and unilaterally, extended the sovereign territory of the United States to 200 miles offshore rather than the 12 mile standard limit demarcating international waters.[82] In 1984 the United States pursued and secured multilateral agreements with the United Kingdom, France, Belgium, Germany, the Netherlands, and Japan to respect each other's effective extensions of territorial waters for the purpose of licensing their respective nationals to carry out seabed mining. While these other countries, and most of the rest of the world, did eventually go on to ratify the UN Law of the Sea Convention, as of 2015 the USA still has not done so — this despite American instigation of a renegotiation of the treaty in 1994 that made it more palatable to all of the other countries that did go on to accept it. While America may eventually sign this treaty, the more than 30 years of resistance and calls for renegotiation are a significant sign of what can be expected when even more valuable lunar resources are at stake and far fewer countries are in a position to vie for access to them.

It is no surprise then that the United States, and all other major spacefaring nations, seem to have no intention of signing the Moon Agreement of 1979.[83] Articles 11 and 18 of this agreement place a de-facto moratorium on any lunar resource exploitation until and unless an international organization structured on a one-nation one-vote system of direct democracy has been established to ensure, by majority vote, that any material benefits afforded by the Moon are shared with everyone on Earth in perfect amity and equality.[84] The "Moon and its natural resources" are declared "the common heritage of mankind" with a specification that strongly prohibits any private ownership of "natural resources in place [on the] … surface or subsurface of the Moon…"[85] The treaty technically entered into force with its requisite ratification by five countries and four others beyond

---

82 Ibid., 290.

83 Ibid., 286.

84 Ibid., 287–288.

85 Ibid., 287.

this minimum threshold, but without the backing of any of the countries that are actually major players in space — among them the Earth's major military powers — there is no possibility for enforcement.

Given the aforementioned illegitimacy of environmental concerns with respect to helium-3 mining on the Moon, it is hard to see what has motivated the signatories other than resentment. Those that have ratified the 1979 Moon Agreement are the nations which, at least at the time, had no possibility of accessing lunar resources themselves and apparently wished to restrain those who are capable from doing so until the latter meet their demand of sharing these resources, on an equal footing, with people who have not worked for them. While it is understandable how, as a leader of the Non-Aligned Movement in 1979, socialist India chose to ratify this agreement, one wonders whether the India of today — with a booming capitalist economy and an increasingly robust space program, will continue to view lunar resources as a cause for resentment or as an opportunity for further development. Broader ratification of the Moon Agreement of 1979 would clearly discourage private investment in lunar-mined helium-3 fusion power, and so Schmitt urges the United States and other spacefaring nations interested in such a venture (China, Russia, Japan) to formally reject the agreement and close the door to any future renegotiation of it.[86]

This brings us, finally, to the Outer Space Treaty of 1967, which is the international legal framework that Harrison Schmitt favors for lunar helium-3 mining. More than 110 countries are parties to this treaty, including all spacefaring nations.[87] Schmitt acknowledges that some have interpreted this treaty to prohibit private property by implication of its denial of claims of national sovereignty over any part of the Moon. However, Schmitt claims that "most objective analysts" do not read the treaty this way and he offers his own very permissive reading of the implications of the treaty's explicit provisions.[88] Article VIII reads: "A State Party on whose

---

86   Ibid., 292.

87   Ibid., 282.

88   Ibid., 282–286.

registry an object launched into outer space is carried shall retain jurisdiction and control over such an object, and over any personnel thereof, while in outer space or on a celestial body." Article XII adds: "All stations, installations, equipment and space vehicles on the Moon ...shall be open to representatives of other states parties on the basis of reciprocity; such representatives shall give reasonable advance notice of a projected visit." Schmitt notes that this need for advance notice in Article XII actually implies a right of exclusivity for each nation over its own facilities which, when taken together with civil and criminal laws provision of Article VIII provides for at least enough recognition of national administrative ability — if not sovereignty — to allow for the licensing and development of private lunar resource exploitation. Furthermore, according to him corporate proprietary information is not endangered by the specific wording of Article XI: "State parties agree to inform... the public..., to the greatest extent feasible and practicable, of the nature, conduct, location and result of their activities... [on] the Moon." This may prohibit, or at least declare illegal, a sovereign nation's installation of clandestine military or intelligence facilities on the Moon, but the commitment to publically share information "to the greatest extent feasible and practicable" does not necessarily require divulging information harmful to the competitive advantage of a private enterprise. At least, that is Schmitt's claim.

Private enterprises will be authorized to operate on the Moon by state parties to the agreement, pursuant to their contractual obligation to abide by its provisions, as indicated by Article VI: "Activities of non-governmental entities... [on] the Moon... shall require authorization and continuing supervision by the appropriate state party." Finally, as already discussed, helium-3 mining protocols in particular will not pose any environmental damage to the Moon, while it will increase the ecological welfare of the Earth, and is therefore not in violation of Article XI: "State parties shall pursue studies of... the Moon ...and exploration ...so as to avoid their harmful contamination. If in doubt, consultation shall be initiated before proceeding with the activity in question."

So on what basis do dissenters argue that the kind of licensed venture Schmitt has in mind ought to be considered illegal under the Treaty?

Three articles are of particular relevance here. Article 1, paragraph 1 of the Outer Space Treaty states: "Exploration and use... shall be carried out for the benefit and in the interests of all countries... and [the Moon and other celestial bodies] shall be the preserve of all Mankind." The next paragraph of this first article goes on to add: "The Moon... shall be free for exploration and use by all states without discrimination of any kind, on a basis of equality and in accordance with international law, and there shall be free access to all areas..." Schmitt interprets paragraph 2 as being permissive in nature; it supposedly allows "private entities, operating under license from a state" to carry out their activities "anywhere on the Moon" in the fashion that international law allows free movement on the open seas. Yet the issue is not free movement, it is resource exploitation and as pointed out above, the Law of the Sea treaty was controversial in this regard even though international waters are not "territory" in the same way as the Moon. When taken together with the first paragraph of the same article, this provision could just as readily mean that nothing on the Moon belongs to anyone in particular but to everyone equally. This is in fact the implication of it that was fleshed out by the subsequent 1979 Moon Agreement.

So what is Schmitt's response to Article 1 paragraph 1? He claims that "for the benefit and in the interest of all countries" can be accommodated insofar as "Uses, such as the extraction of resources" will be "for broad-based use on Earth to raise living standards..." And how on Earth is that going to be guaranteed? Listen to this:

> A lunar private initiative's principal business objective would meet this guideline [of being "for the benefit and in the interests of all countries"], namely, to provide the energy, economic and environmental benefits of lunar resources to all customers without geographic or national discrimination other than those required by law. Further, once retained earnings reach an appropriate level, the possibility will exist for shareholder approval of the endowment of a foundation for the purpose of facilitating global access to the benefits of helium-3 fusion.

The first question is "required by what law"? The law of the nation licensing the corporation, which may have enacted a unilateral or multilateral trade embargo of any number of nations for whatever reasons it sees fit.

Moreover, how could Schmitt possibly imagine that the intent of Article I paragraph I is consistent with a corporate board of a private enterprise setting up a charitable foundation for those nations so benighted as not to have any access to either the Moon or fusion power technology?

Most significantly, Article III flat out asserts that: "The Moon ...is not subject to national appropriation by claim of sovereignty, by means of use or occupation, or by any other means." Regarding the last of these, Schmitt draws an analogy between private property rights over mining resources in the United States which, when these are obtained on public lands, extend only to the resource acquired and the equipment by means of which it was exploited *not to the public lands themselves*.[89] He acknowledges that this fine distinction is lost on much of the rest of the world, where mined minerals are considered a natural resource owned by the state in the same fashion as the lands from which they were extracted. Are we to assume that only liberal free-market capitalist countries will be mining the Moon for helium-3? Schmitt seems to take this for granted when he claims that Articles I and II imply that no lunar resources can be acquired by a nation for the exclusive use of its own population: "A lunar private initiative would market lunar resources within the global marketplace..."[90]

But there is a bigger problem with this reading of Article III. The analogy to mining rights in the United States is a poor one. The "public lands" are not owned by the mining corporation because they already belong to the United States of America, and only on that basis can the government of this sovereign nation lease mining rights to a private interest and acknowledge private ownership of the mined resources. Private property is protected by a sovereign authority and, as Schmitt himself acknowledges: "State or public ownership of mineral bearing lands has its roots in ancient, sovereign rights — that is, the rights of the crown."[91] On the Moon, then, who is the sovereign with the power to license operations on "public

---

89 Ibid., 281.

90 Ibid., 284.

91 Ibid., 281.

lands" or recognize the resources acquired thereby as private property of its citizens?

International law on Earth works because the Earth's habitable sphere is divided into sovereign territories that enter into agreements with respect to one another's territoriality. International waters are a false analogy to the Moon. No recognized nation has citizens living in the Oceans — at least not yet, and if any members of the international system ever have significant submarine colonies they will face the same problems of sovereign jurisdiction as we are faced with on the Moon. At one point Schmitt even envisions a lunar colony or colonies developing to a magnitude sufficient for a declaration of "political and corporate independence from Earth".[92] If Schmitt is willing to consider, and even to recognize, this form of sovereignty over the Moon's territory then he ought also acknowledge that any substantive colony of permanent settlers with their own families represent not just an extension of their culture and religion to the Moon, but also an extension of the expectation of sovereign protection of their lives and property. The moment any kind of securing of a living and working space against violent assault by a potential enemy takes place, a claim of sovereignty is being extended over a territory. In other words, the impending energy crisis forces the question of world government upon us whether or not we are ready for it. Considered as an energy resource, the Moon is indeed the Persian Gulf of the 21st century. Whoever seizes control of this powerhouse will also effectively be ushering in a new geopolitical order.

---

92 Ibid., 292.

CHAPTER 6

Aryan Imperium (*Iran-Shahr*)

By now we have seen that within a timespan not longer than a single generation, one world government needs to be established. This world state will be tasked with deciding on, and enforcing, uniform worldwide regulations that allow us to benefit from the many promises of biotechnology while avoiding some of the most monstrous perils of cloning and genetic engineering. The same world state must guide us into a post peak-oil industrial economy, preventing the specter of war in space for control over the Persian Gulf of the 21st century.

We have also seen that a bureaucratic world state will not suffice. Certain developments in robotics mean the end of personal privacy and even a modicum of security within what we have hitherto taken to be our most private spaces. As we live ever more of our lives in cyberspace, identity theft is coming to have a much more literal meaning. All in all, the convergent technological advancements that we have looked at require a maximal trust society simply for the sake of human survival. We need a world society with total interpersonal transparency, bound together by entirely sincere good will. The analysis of the first two chapters has shown why the modern political paradigm of human rights and liberal democracy is hopelessly impotent in the face of such a challenge. The only way that an organic state could come into being on a global scale within the next 30 years is on the basis of an already existing ethos, a living tradition that is inter-civilizational and global in scope.

There is one and only one contender for this: the common Aryan heritage of the Indo-European civilizations. In his book, *The Indo-Europeans*,[1] Alain de Benoist returns to one of the foundational elements of his magnum opus, *View from the Right*.[2] He reviews the vast body of literature produced by tens of scholars over the past three centuries on the subject of the common ethnic and linguistic roots of the European, Iranian, and north Indian peoples. Emanating from a homeland that it seems increasingly likely was located somewhere between Ukraine and the Caucasus, or between the Black Sea and the Caspian Sea, a single ethnic group speaking a single language branched out in westward and eastward migrations in the course of which they gradually became differentiated from one another. Greek, Latin, German, Persian, Sanskrit, and the numerous later languages that have evolved from these classical languages, share deep structures and numerous cognates in their vocabulary that attest to their having branched out from a single trunk that scholars refer to as Proto Indo-European. Interesting attempts have been made to reconstruct this root language, which was spoken as late as 4,000 BC.[3]

The various Indo-European civilizations were all world-colonizing, whether in a military sense or in a cultural one. Taken together they eventually brought our entire planet under their dominion, with the two most vast colonial structures being the Persian Empire and British Empire. The former was the largest Indo-European superpower in terms of population (including nearly half of Earth's denizens at its zenith), and the latter was the most extensive colonial realm in geographical terms. Buddhism's conquest of East Asia, from its cradle in northern India, can be seen as a purely cultural Aryan conquest, one which, as we will see, was primarily carried out by Iranian missionaries traveling the Silk Route. The Indo-Europeans originated nearly all of the exact sciences and the technological innovations based on them, the rich artistic and literary

---

1 Alain de Benoist, *The Indo-Europeans: In Search of the Homeland* (London: Arktos, 2016).

2 Alain de Benoist, *View from the Right, Vol. 1* (London: Arktos, 2017).

3 Carlos Quiles and Fernando López-Menchero, *A Grammar of Modern Indo-European: Prometheus Edition* (2012).

traditions of Europe, Persia, and India, as well as major philosophical schools of thought and religious traditions such as Platonic and Germanic Idealism, Enlightenment Progressivism, Zoroastrianism, Hinduism, and Buddhism. This heritage is exclusionary, as Carl Schmitt rightly understood the constitution of all properly political states to be. It is, however, by far the most broadly encompassing basis for the emergence of a world state from out of the world war of civilizations and in the face of the technological apocalypse.

In his analysis of the transition from the era of international conflict to the epoch of the clash of civilizations, one mistake that Samuel Huntington makes is to consider Iran part of a so-called "Islamic Civilization" that includes Arab, Turkic, and Southeast Asian elements. Islam's allegedly civilized veneer is a product of its parasitic invasion of Iranian Civilization, which is a distinct world-historical civilization. Moreover, Iran is not just one great civilization amongst a handful of others, it is that crossroads of the world that affords all of humanity the possibility for a dialogue of civilizations toward the end of a new world order based on something other than the destruction of distinct cultures and the deracinating unification of all peoples and nations on the basis of the lowest common denominator. A handful of ideas or ideals integral to the structure of Iranian Civilization could serve as constitutional principles for an Indo-European world order: the reverence for Wisdom; industrious innovation; ecological cultivation; desirable dominion; chivalry and tolerance.

As I understand it, a civilization is a super-culture that demonstrates both an internal differentiation and an organic unity of multiple cultures around an ethno-linguistic core, one which roughly corresponds to Alexander Dugin's concept of the *narod*.[4] For example, the Mandarin-speaking Han Chinese are the ethno-linguistic core of the *narod* of Chinese Civilization, although it includes Cantonese, Manchurian, and even Mongolian cultures. Classical Rome and its Latin language remain the core of the Western *narod*, despite the increasingly significant role played by the Germanic elements of Western Civilization. For the past

---

4   Alexander Dugin, *The Rise of the Fourth Political Theory* (London: Arktos, 2017).

2,500 years the Persian ethnicity and language has been the core of Iranian Civilization, which also includes other ethnically and linguistically Iranian cultures. Kurdistan, Azerbaijan, Ossetia, Tajikistan (including the Tajik parts of Uzbekistan and Afghanistan), the Pashtun territories, and Baluchistan are as much a part of a distinct Iranian Civilization as Spain, England, France, Germany, and Sweden are part of a Western Civilization that has its origins in Greece and Italy.

However, the boundaries between world civilizations are never clear-cut and Western Civilization overlaps with Iranian Civilization in southeastern Europe, since Ukraine is an originally Iranian nation, the homeland of the Scythians, and Sarmatian tribes were once dominant in Bulgaria and Croatia, also lending these European territories roots that trace back to Iran more authentically than to the rest of Europe. The roots of the greatness of European civilization are inextricable from the ancient Zoroastrian heritage of Iran. The idea that the Persian Wars represented some kind of clash of civilizations, let alone a race war, is totally anachronistic and delusional.

Ancient Iran was the first and greatest white colonial empire, counting nearly one out of every two persons inhabiting the Earth among its subjects. Its Caucasian ruling class of Persians and Medes (Kurds) was racially identical to the various ethnicities of Europe. The Greco-Persian wars were white on white violence, like later wars that the Romans fought with the German or Celtic "barbarians" long before the idea of a united "Europe." Iran was opposed not by a unified Greek or European civilization — neither yet existed — but by a loose alliance of Greek city states. Many other Greeks sided with Iran.

For six centuries, from at least the Trojan War of Homer's *Iliad* in 1230 BC to Hesiod's *Theogony* in 650, we see essentially no change in the mythic world-view of the Greeks. Then suddenly in the 6th century BC we have 12 "philosophers" appearing within just one century. It is more than a strange (and suspiciously overlooked) coincidence that the sudden rise of philosophy like a meteor from a Greek mind sunken for millennia in the dark marshes of fatalist myth and superstition, coincides exactly with the

Persian conquest and colonization of Greece beginning in the 6th century BC and enduring for well over a hundred and fifty years.

The reverence for Wisdom was at the core of Zarathustra's teaching. The most fundamental constitutional principle of the Indo-European worldview is the supreme value that it places on the reverence for, the pursuit of, and the adherence to, Wisdom. Zarathustra speaks of *Ahura Mazda* or the "Titan of Wisdom" as the deity who is supremely worthy of worship. Sometimes he simply refers to this God as *Mazdâi* or "Wise One." The opening lines of Ferdowsi's *Shahnameh* are, *be nâme khodâvandé jân-o-kherad* or "in the name of the Lord of life and wisdom."

Contemplation with the aim of understanding, the acquisition of knowledge, or the discernment of truth, only make any sense on the presupposition of a cosmic order like that suggested by the Persian idea of *Ordibehesht* or *Ârtâ Vahishtâ* in Zarathustra's Avestan dialect. Implicit in the idea of *Ordibehesht* is another concept that entered the West via Iranian influence, the concept of a concept as such. One significant aspect of the Iranian notion of cosmic order is that ideal forms or archetypes of all things, called *fravashis*, preexist and shape physical instantiations of them. This was the first appearance of such a proto-mathematical and purely rational or abstract way of thinking. It influenced Plato by means of his membership in the Pythagorean Order. The crystalline forms are often imagined as abiding in a realm of light, which is distinct from an abyssal darkness of chaos full of unshapely vermin. If the latter were not a distinct dimension of existence, *Ahriman* would have no stronghold or staging ground from which to derange the physical world and militate against the influence of the archetypal reality. It is important to note that even the creative intellect of *Ahura Mazda* is constrained by these archetypes integral to *Ordibehesht*. Cosmic order is not the product of an arbitrary divine decree. *Mazdâ* himself contemplates the pure forms. In this sense, although *Mazdâ* is a Creator whose creative genius is the highest expression of *Soroush* or Inspiration, *Mazdâ* also has a rational mind.

In the Aryan worldview a truly human life is one devoted to the cultivation of *Bahman,* the best thinking or intellectual excellence that can discern and bring one into harmony with the cosmic order so that one may contribute to the great work of the creative spirit. *Bahman* is a contraction of the Avestan *Vohuman*, which could also be contracted in the form of *human*. This is the prototype of the Latin concept of *humanitas*, which is either an Indo-European cognate or found its way into European civilization through Iranian influence — possibly via Mithraism. The idea is that one is only a properly human being if one cultivates ones mind. As Darius' would-be court philosopher, Heraclitus of Ephesus says "It belongs to every man to think well…" Interestingly, although our intellectual faculty is what separates us from animals, *Bahman* is symbolically associated with the animal kingdom. I think that this is because how we treat animals is a reflection on our own humanity. What separates us from them also makes us responsible for their welfare. There were extremely severe laws against cruelty to animals in the first Persian Empire. For example, mistreating a dog was as grievously punished as abusing a human child. In northern India, under the reign of the Kushans, who were Scythian converts to Buddhism, there were even more wide ranging laws against cruelty to animals.

The idea that being ever more firmly rooted in the right order also yields an ever increasing serenity, or *Sepandârmad*, is another fundamental idea of Zarathustra's worldview. In the Yoga tradition of India, and the Buddhist psychology that developed from out of it, this is referred to as *samadhi*. The Greek philosophers, such as Plato and Aristotle, refer to it as *sophrosyne*. In the older Iranian tradition, this abiding calm that comes over a contemplative and conscientious person is associated with the element of Earth, and in this regard it is one of three closely related principles that demarcate the ecological dimension of Zarathustra's message. Among the Bounteous Immortals of Zoroastrianism, *Spentâ Ârmâiti* in Avestan or *Sepandârmad* in middle Persian is depicted as a feminine figure — a kind of Mother Earth goddess.

Closely related to her is *Khordâd*, the spirit of health or wholeness, which is associated with the element of water. The Iranian cult of Anahita, the Lady of the Lake and Virgin mother of Mithra — who gives birth to him on the long night of the Winter Solstice — included baptism rituals with holy water that were intended to make one hale and confer a cleansing spiritual perfection. The idea here is not only that health follows from the intellectual discipline of *Bahman* and the serenity of *Sepandârmad*, so that a disordered mind and a volatile life are unhealthy, but also that proper attention to bodily health and well being is a prerequisite of success in seeking enlightenment.

Iranians conceive of the attainment of Enlightenment in terms of *Amordâd*, or *Ameratât* in Avestan. This is often translated as "immortality" but it literally means un-deadness, in the sense of vitality. Intellectual excellence, justice, chivalry, serenity, and health ultimately lead one to this superhuman or supremely human condition. Zarathustra's metaphysics conceives of this in terms of an alchemical transformation of the human condition that takes place at the end of history. This teleological — rather than cyclical — conception of time and world ages was eventually adopted by Germans such as Schelling and Hegel, who become expositors of its most developed version. On the whole, *Sepandminou* works to progressively and innovatively improve the human condition throughout successive epochs. After a final apocalyptic conflict, referred to both as "the great event of choice" (namely of choice between the two spirits) and also as the *Frashgard* or the "renewal of existence", all of those who have chosen rightly to be champions of Truth and Justice attain their archetypally perfect form. This *farvahar* acts as one's guardian angel during one's life, enjoining us to become who we are. Interestingly, in line with the reverence for women in the Indo-European heartland, the exteriorization of one's inner conscience or *daenâ* that one embodies after the apocalypse is feminine in form, whether one is now a man or a woman; she looks like a Valkyrie.

But *Amordâd* does not only refer to personal immortality. It also has another meaning that connects it more closely to the aforementioned ecological principles of Zarathustra's message in order to form the trio of serenity, health, and vitality. In this sense *Amordad* is associated with the element of vegetation — the lush greenery of trees, plants, fruits, and vegetables. It was considered a sacred duty to propagate agriculture so as to participate in the divine creation and make the living world more bountiful.

This attitude had real consequences in terms of the kind of industrious innovation that one is enjoined to embrace in order to further *Sepandminou* or the Spirit of Innovation and Development. Iranians invented a technology known as the *Qanât* in order to channel water across hundreds of miles so as to be able to make gardens bloom in the middle of deserts. A system of tunnels cut through rock, accessible from the surface through wells, and leading to dams and distribution channels, would act to artificially raise the water table and conduct water from aquifers to land that would otherwise be impossible to irrigate. The gardens that were created as a consequence of this engineering marvel were referred to as *Paridaezâ*, which is where the word "paradise" in the European languages originally comes from.

The concept of *Paridaezâ* can be seen as a metaphor for the governing philosophy of the ancient Persian Empire, and the geopolitical vision that ought to be adopted by the Indo-European World State. One can have a variety of flowers and crops growing in one's garden, and they are beautiful in their diversity. But in order to guard this variegated beauty, any hybridizations that take place must be well considered and carefully crafted. Most importantly, the garden has to be weeded, and harmful pests that threaten everything growing in there have to be eradicated. It is not difficult to imagine, in today's political geography, who falls into the category of such weeds and pests. At any rate, the construction of *Qanâts* is only one example of the Aryan civilizational idea of *Âbâdsâzié Giti*—the industrious beautification of the world.

This brings us to the legal and socio-political philosophy of the Indo-Europeans. It is radically utopian. The core concept of relevance here is *Shahrivar*, a middle Persian contraction of the Avestan *Khashatrâ Vairyâ* or the Desirable Dominion. It could also be translated as the Most Choiceworthy Kingship. In other words, it is the ideal form of government, that which one would choose if only one could see it. *Shahrivar* comes into being when *Ordibehesht* is not only discerned by *Bahman* on an individual basis and embodied by a single person, but when an entire political order is rightly guided to bring society as a whole into harmony with cosmic order and the creative divine intellect. *Shahrivar* is elementally associated with metal. This has alchemical significance. Imagine the metal sword of the just ruler being forged in the fire of *Ordibehesht*.

Beginning with Zarathustra's alliance with the Shah Goshtasp and continuing on through the partnerships of Darius and Heraclitus, Shapur and Mani, Kavad and Mazdak, Khosrow and Bozorgmehr, the willingness to be advised by a philosopher or a council of sages was one of the signs of the rightly guided ruler who has the *farré kiâni* (ancient Persian *Xvarnah*). This is the kingly glory symbolized by the originally Iranian artistic convention of the solar halo around the head — which entered European art through the spread of Mithraism and Asian art through the Iranian invention of Mahayana Buddhism.

When the Persian army crossed the Hellespont into Greece it was as a sword in the hand of a leadership concerned with the propagation and prospering of Zarathustra's thought-provoking message. The empire founded by Cyrus, organized by Darius and fostered by Xerxes — was not only Zoroastrian in its society and culture but was actively functioning as an embodiment and missionary of Zarathustra's doctrine. Let us look at Achaemenid society as described by the Greek historian Herodotus, who encountered it first hand. Regarding the religion of the common Persians he writes: "They are not wont to establish images or temples or alters at all; indeed, they regard all who do as fools, and this, in my opinion, is because they do not believe in gods of human form, as the Greeks do."[5] He adds that the Persians do not believe in a God so petty as to entertain prayers asking for an alleviation of the particular problems of any given individual, and so their only lawful 'prayer' is for the well-being of all.

Herodotus goes on to explain that the highest value and principle around which their society turns is "truthfulness" and contempt for deceit. We find in his account evidence of an active implementation of Zarathustra's principle that thoughts, words, and deeds must perfectly reflect each other: "Whatsoever things it is not permitted to them to do, of these they must not even speak. Lying is considered among them the very basest thing and, second, indebtedness…because, as they say, a debtor is

---

5 Herodotus, *The Histories* (Chicago: University of Chicago Press, 1987), 1.131.

bound to lie somewhat."⁶ Apparently from the age of five and up, Persian children were rigorously disciplined to make a practice of always telling the Truth. For Herodotus writes: "They train their sons from their fifth to their twentieth year in three things only: horsemanship, archery, *and truth-telling*."

One particularly colorful practice which reveals the love of Truth in Achaemenid society is that, according to Herodotus, the Persians would never enter into debates and discussions of serious matters unless they were drunk on wine. The decisions arrived at would later be reviewed in sobriety before being executed:

> They are very addicted to wine...[and]...They keep very strictly to this practice, too: that they are wont to debate their most serious concerns when they are drunk. But whatsoever they decide on, drunk, this the master of the house where they are when debating proposes to them again on the next day, when they are sober. And if they like it, too, when sober, they act on it; but if they do not like it so, they let it be. And whatever they debate, in preliminary fashion, sober, they give to final decision drunk.⁷

It seems that they believed the wine would embolden them to drop all false pretenses and get to the heart of the matter.

Yale philologist Stanley Insler has also noted a curious feature of Achaemenid society which testifies to its wholehearted embrace of Zarathustra's principles. Ancient Persian names were always descriptive of a person's qualities and would be chosen by parents as a wish for the kind of person they would like to see their child become. We have an immense inventory of 1,500 such names inscribed on the many Old Persian tablets surviving from the period. Some of them are: *aspâugurâ* — "strong as a horse"; *hubâoidi* — "sweet-smelling"; *virâka* — "little hero"; or *vsavâh* and *humizdâ* — "having good fame" and "winning a good prize". These express longings for strength, heroism, beauty, fame and fortune. But what is striking is that the vast majority of names on these tablets do *not* refer to such qualities, bur rather incorporate the attribute of "Truth" or *Arta* in

---

6   Ibid., 1.138.

7   Ibid., 1.133.

Old Persian (*Asha* in the even older dialect of Zarathustra), for example we find: *Ârtâpanâ* — "Protector of Truth"; *Ârtâkâmâ* — "Lover of Truth"; *Ârtâmanâh* — "Truth-Minded"; *Ârtâfarnâh* — "Possessing the Splendor of Truth"; *Ârtâzustâ* — "Delighting in Truth"; *Ârtâstunâ* — "Pillar of Truth"; *Ârtâfridâ* — "Prospering the Truth"; *Ârtâhunârâ* — "Having the Nobility of Truth", and so forth.[8]

The political leadership of the Achaemenid dynasty upheld the Zoroastrian principles at the core of Persian society from its very founding. When Cyrus the Great invaded Babylon and deposed the brutal King Nabonidus he declared the world's first humanitarian charter as the inaugurating seal of the Persian Empire. It reads in part:

> ... I am Cyrus. King of the world. When I entered Babylon... I did not allow anyone to terrorize the land... I kept in view the needs of its people and all its sanctuaries to promote their well-being... I put an end to their misfortune. The Great God has delivered all the lands into my hand; the lands that I have made to dwell in a peaceful habitation...[9]

Cyrus' successor, Darius, is remembered for inventing an ingenious system of organization for his vast Empire, which is the direct predecessor of the federation system of states and governors (*Satrapies*) employed by the United States. Thomas Jefferson studied Xenophon's *The Education of Cyrus*, which holds up the first Persian Emperor as the ideal statesman.[10] A great effort was made never to ravage a conquered territory, and even if possible, to keep its indigenous king in power and only append to him a Persian governor who would assure the upholding of basic human rights, the collection of taxes and the supply of young men for the Persian army. Cyrus spared the lives of all three kings of the major kingdoms he conquered: Astayages of Media, Nabonidus of Babylon, and Croesus of Lydia. Even though the latter had attacked Persia first, and without provo-

---

8   Stanley Insler, "The Love of Truth in Ancient Iran" in *An Introduction to the Gathas of Zarathustra* (Pittsburgh, 1990).

9   Roland G. Kent, *Old Persian: Grammar, Texts, Lexicon* (American Oriental Society, 1953).

10 Xenophon, *The Education of Cyrus* (Cornell: Cornell University Press, 2001).

cation, Cyrus made him an advisor at the Persian court.[11] This stands in striking contrast to the contemporary Greeks' routine of plundering cities conquered in battle and raping their noble women, as well as the way in which the Assyrians would raze a city to the ground and bind its people into slavery. Not only were occupied territories spared the pillage typical of conquests of the time but great projects of restoration were undertaken. Such was the state of affairs that the people of many oppressive kingdoms greeted the arrival of the Persian army enthusiastically as it made its way from India and Western China to Egypt and finally, Greece.

In short, though the reign of the Achaemenids is often recognized as the world's first real 'Empire' — it is in fact more appropriate to call it a *Universal State* — with its decentralized system of local governors, vast royal roads, and the world's first postal system. Such a designation is more fitting, above all because of the sense of humanistic cosmopolitanism which the Achaemenid dynasty fostered. Herodotus writes: "The Persians welcome foreign customs more than any other people."[12] He explains that they adopted whatever they saw as *universally* best in and of itself, its ethnic origins did not matter. Achaemenid art is also recognized as a unique attempt to consciously mix the artistic traditions of *all* the subject people into a humanist style that would reflect their new unity. Native custom was far less important than perpetual reflection on the Good.

By comparison to the history of kingship and conquest before them, the Achaemenids were not pursuing a narrow-minded nationalistic agenda of subjugation. They were seeking to secure peace upon the face of the earth through the liberation and prospering of all its children. They were the first people in recorded history to envision "humanity" as an abstraction set over and against tribal or ethnic identity. Harvard Professor Richard Frye writes in his *Heritage of Persia*:

> In the victories of the Persians... what was different was the new policy of reconciliation and together with this was the prime aim of Cyrus to establish a

---

[11] Reza Zarghamee, *Discovering Cyrus: The Persian Conqueror Astride the Ancient World* (Washington: Mage Publishers, 2013), 102–110.

[12] Herodotus, *The Histories*.

pax Achaemenica..... If one were to assess the achievements of the Achaemenid Persians, surely the concept of One World .... the fusion of peoples and cultures was one of their important legacies.[13]

In this we see a patronage of the *stewardship of the earth* that lies at the heart of Zarathustra's doctrine. The Achaemenid openness to reflecting on the inherent good of others' traditions vs. their own, and the like, is a reflection of a people whose God is not 'the God of the Persians' (as that of the Jews is emphatically "the God of Israel"), but the God of the whole Living World. We hear the following of the Achaemenid justice system in practice from Herodotus:

> ...no one, not even the Great King himself, may kill anyone on charge of a single crime, nor may anyone of the rest of the Persians do irremediable harm to any of his servants on occasion of a single act. Only if, on consideration, he finds the wrongdoings more in number and greater than the good deeds may he use his pleasure.[14]

In one inscription of Darius the Great at Persepolis in Iran, he sees "the Lie" as equally devastating to the land as invasions and famine and prays to *Ahura Mazda* to allow his people to abide in Truth: "Darius the King says: May Ahuramazda bear me aid...and may Ahuramazda protect this country from a (hostile) army, from famine, from the Lie! Upon this country may there not come an army, nor famine, nor the Lie; this I pray as a boon from Ahuramazda..."[15]

Another of the inscriptions that Darius left us at Persepolis rejects both the oppressive doctrine of might makes right and the politics of *ressentiment* that passes for 'social justice' today: "By the favor of Ahura Mazda I am of such a kind that I am a friend of the Right, and not a friend of the Wrong; it is not my desire that the weak man should suffer injustice at the hands of the strong, *nor is it my desire that the strong man*

---

13 Richard Frye, *The Heritage of Persia* (Costa Mesa: Mazda Publishers, 1993), 102.
14 Herodotus, *The Histories*, 1.137.
15 Roland G. Kent, *Old Persian: Grammar, Texts, Lexicon*.

should suffer injustice from the weak."¹⁶ On this view, democracy is most certainly a tyranny of the majority. Nor should the aristocracy of a proper Aryan imperial government be confused with oligarchy, or the tyranny of bankers and merchants, who often use the disorder of a democracy to manipulate the ignorant masses. Aristocracy is the rule of the best people, meaning the wisest and most cultivated people in accordance with the ideal of *Bahman*, not the rule of the wealthiest people or those preoccupied with material gain.

In fact, a real concern with socioeconomic welfare is central to the conception of ideal governance developed by advanced Aryan societies against the backdrop of an archaic caste system that was rejected both by Zarathustra and the Buddha, and that was radically modified into a meritocracy by Plato. While it is also characteristic of the social ethics of Buddhism, even as it spread deep into East Asia, *Daheshmandi* or charitableness is most characteristic of the Aryan political order within the Persianate world. There is no difference whatsoever between a genuine meritocracy and a true aristocracy, of the kind that Plato and Pythagoras advocate against Athenian democracy and on the basis of their familiarity with Persian imperial culture. Herodotus tells us that aside from deceit, Persians saw indebtedness as the greatest evil — because "a debtor is bound to lie somewhat."¹⁷ Poverty and indebtedness have always been viewed by Iran's just rulers as the breeding grounds of criminality and deceitfulness. Consequently poverty ought to be abolished. A just society is one with an economic system that organizes the production of goods and the distribution of resources in such a way that not a single person is poor, and every citizen is guaranteed a standard of living that affords them the possibility of cultivating human excellence to the maximum given their own potential.

Darius the Great assert that his governance is based on *Ordibehesht* and *Bahman*: "I desire what is right. I am not a friend of the man who follows the Lie. I am not hot-tempered; the things that develop in me during a dispute I hold firmly under control through my mind, I am a firm

---

16 Ibid.

17 Herodotus, *The Histories*, 1.138.

ruler over myself."[18] Such claims are essentially echoed in the Behistun inscriptions. These are no idle words for we know that the Achaemenid rule was the first and only Empire in history not to employ slavery but to outlaw it, no doubt because in Zarathustra's view to live as a slave would preclude the possibility of developing a cultivated intellect (*Vohuman*) so as to truly become a *human* being. Every worker at Persepolis was paid a living wage.[19]

Persian influence on Greek thought began with Cyrus' invasion and occupation of Lydia, whose capital city, Sardis, was according to Herodotus "the resort of all wise men of Hellas."[20] A short time later Lampsacus was one of the first Hellenic towns to be conquered by Cyrus and once under the authority of the 'pax Achemenica' it became a haven of thinkers persecuted for deviating from tradition in Greek mainland cities such as Athens. The channels for influence increased drastically when by 450 BC Darius had extended Persian rule beyond the Hellespont to the shores of the Danube in the North, and the Adriatic sea to the West. Herodotus reports that Darius' conquest brought many Zoroastrian colonists with it to settle in Greece, particularly in Macedonia and Thrace in cities such as Abdera and Eion. We know of instances in which Zoroastrian Magi became the tutors of children of Greek aristocrats, one such case being that of Protagoras, whose father Maendrius welcomed and feasted Xerxes.[21]

Alfred North Whitehead once famously remarked that the whole history of Western Philosophy consists of a series of footnotes to Plato. The two greatest influences on Plato were Heraclitus and Pythagoras. Heraclitus has a very explicit connection to Iran. He was at one point invited by Darius the Great to become *the* Court Philosopher of the Achaemenid dynasty. If we look at the remaining *Fragments* of Heraclitus'

---

18 Roland G. Kent, *Old Persian: Grammar, Texts, Lexicon.*

19 F. Altheim and R. Stiehl, *Die aramäische Sprache unter den Achaimeniden* (Frankfurt, 1961), 173.

20 Herodotus, *The Histories.*

21 Ruhi Afnan, *Zoroaster's Influence on Anaxagoras, the Greek Tragedians and Socrates* (New York: Philosophical Library, 1969), 71–72.

writings on Nature, we see many parallels between his thinking and that of Zarathustra and these have no precedent whatsoever in Greek thought.

One of them is the reverence for Wisdom and "thinking well" set apart from all else. In fact, at one point Heraclitus cautiously hints that Zeus is not the true God and he refers to the true Lord as "the Wise One."[22] He takes the chief aim of human life to be the cultivation of the best or most intelligent thinking, and to align one's thoughts, words, and deeds: what you say and do should be based on careful contemplation, not a casual unreflective acceptance of what others have said.[23] In this regard, Heraclitus levels a scathing critique at the ritual priesthood and at the poetic bearers of tradition or custom among the archaic Greeks.[24] He is as critical of Homer and Hesiod, and of the ritualistic priesthood of his society as Zarathustra is of the priestly caste repeatedly targeted throughout the *Gathas*.[25]

Heraclitus adopts fire — an undying or everlasting fire — as the symbol of cosmic order. This idea of cosmic order, which he refers to in terms of the interpenetration of *cosmos* and *logos*, is identical to Ashâ or Ârtâ in Persian thinking, which is associated with the element of fire in the *Gathas* of Zarathustra.[26] This metaphorical eternal fire of Lord Wisdom's mind becomes the central sacred symbol of Zoroastrianism. Such fires are perpetually tended at Zoroastrian temples to this day. Heraclitus also lays an emphasis on dualistic or oppositional forces as the wheelwork of evolutionary development in the cosmos.[27] There are small details which are also significant. For example, one of the *Fragments* refers to throwing out corpses as quickly as one can.[28] This was anathema to the Greek practice of mortuary rites, but is very similar to the Zoroastrians taking

---

22 Charles Kahn, *The Art and Thought of Heraclitus* (Cambridge: Cambridge University Press, 1999), 41, 55.

23 Ibid., 29–37.

24 Ibid., 33, 37–39.

25 Ibid., 81–83.

26 Ibid., 45, 47.

27 Ibid., 63–67, 85.

28 Ibid., 69.

their dead bodies to the *dakhme* enclosures where they would be picked clean by vultures.

Darius extended his invitation to Heraclitus at the moment when the Athenians orchestrated a revolt of the Ionians against Persian rule. The Persian satrap and governor of Ephesus, a man by the name of Hermodorus, was a personal friend of Heraclitus. In one of his *Fragments* Heraclitus goes to the extent of saying that his fellow countrymen who revolted against the Persians ought to be executed to the last man and the city should be left to the children (those not yet brainwashed into opposing enlightened Persian rule).[29] He was an ardent opponent of democracy, which he saw as the mob rule of the ignorant rabble. A man who says something like this, and always means what he says, cannot be considered a 'Greek' other than by birth — unless he is also to be considered a traitor. Rather, by seeing his fellow Ephesians as traitors to a noble or truly Aryan regime sincerely aspiring to be in line with cosmic order, Heraclitus identifies himself as a proud Iranian citizen.

Instead of accepting the invitation of Darius he sequesters himself in the Temple of Artemis. The significance of this has hitherto been lost on commentators. Who is Artemis? Greek historians and anthropologists tell us that she was the goddess of the Amazons, and recent research has demonstrated that the Amazons were a historical people — they were the female warriors among the Sarmatians, an Iranian tribe from the Caspian Sea region that, together with another closely related Iranian tribe, the Scythians, extended their dominion north around the Black Sea and down into the Bosphorus region, where legend has it that they built the original Temple of Artemis as one of the seven wonders of the ancient world.[30] "Artemis" has no clear sense in Greek, but if you read it through the lens of Persian linguistics you get a contraction of the compound *Ârtâ Ameshâ*. Recall the many compound ancient Persian names that include *Ârtâ* or "Truth." The *Ameshâ* from this one is the same as in Zarathustra's *Ameshâ Spentâs*, in other words Artemis means "Immortal Truth" or "Truth, the

---

29 Ibid., 55–59.

30 R. Brzezinski and M. Mielczarek, *The Sarmatians* (Oxford: Osprey Publishing, 2002).

Immortal" — a hypostatization of *Ashâ* somewhat in the style of the later European "Nuda Veritas." The chief symbol of Artemis is the archer's bow and arrow, the ancient Persian symbol of Truth referenced by Nietzsche when he puts this maxim into the mouth of his returned Zarathustra: "To speak the truth and shoot well with arrows, that is Persian virtue."[31]

While Heraclitus did not make it to the royal court of Iran, choosing instead to stand his ground and fight for *Ârtâ* at the furthest Western frontier of the Empire, the other greatest influence on Plato, Pythagoras of Samos, did spend a decade in the capital of the Persian Empire. Pythagoras had traveled from his native island of Samos to Egypt, where he was studying in the temples with the Egyptian priests, when Cambyses II, the son of Cyrus the Great, colonized Egypt and integrated it into the Persian Empire. He gave orders for Pythagoras to be "arrested" and brought to the capital of the Persian Empire. What the Shah meant by this joke was that since the Persians were such seekers of Wisdom (*Mazdâ*), they wanted to bring this curious fellow back to Babylon so that they could pick his brain and see what he could possibly also learn from them. According to Iamblichus and Porphyry, Pythagoras spent twelve years studying under the Magi (*Moghân*) or Zoroastrian priests in the administrative capital of Iran.[32] He returns to his native Samos only when the island is conquered by the Persian Empire. Samos comes under Persian rule in 522 BC and Pythagoras returns there in 520 BC. He spends about a decade at home before moving on to southern Italy, where he establishes a very revolutionary school of thought.

Now we have moved back in time before the generation of Heraclitus and Darius. There was, as of yet, no such thing as "Philosophy" in Greece at all. In fact, it is well known that Pythagoras was the first person in Greece to refer to himself as a "philosopher". He coins the term. In Greek there are a number of different words for love. *Eros* refers to erotic love. *Agâpe* refers to compassion. *Philiâ* in particular refers to friendship. By saying that he is

---

31 Friedrich Nietzsche, "Ecce Homo" in *Basic Writings of Nietzsche* (New York: The Modern Library, 2000), 783–784.

32 Kenneth Sylvan Guthrie, *The Pythagorean Sourcebook and Library* (Grand Rapids: Phanes Press, 1988), 61.

a beloved Friend of Wisdom, or has a relationship of intimate friendship towards Wisdom (Greek *Sophia*, Persian *Mazdâ*) he is evoking that very unique relationship to God that is anathema to ancient Greek religion and even to subsequent Abrahamic faiths (except insofar as their mystical offshoots have come under Persian influence). In Zoroastrianism there is the notion that Man is a Friend of God and that God requires the friendship of Man in order to unfold the plan of righteousness in the world. It is not a coincidence that Philosophy only emerges in Greece after the first man to conceptualize it spends a decade in the capital of an Empire whose religion (Zoroastrianism) is natively and properly referred to as *Mazdâ Yasnâ* or "Wisdom Worship", a doctrine which conceives of the rapport between humanity and Lord Wisdom as an intimate friendship.

Porphyry and Iamblichus tell us that although Pythagoras absorbed a number of influences from a variety of cultures, the Magi had the deepest influence on the fundamental spiritual orientation of the Pythagorean Order. One example of this is Pythagoras' teaching against animal sacrifice and cruelty towards animals. Some folktales suggest that Pythagoras reached India, but there is no good evidence to support his having gone any further East than the capital of the Persian Empire. Consequently, rather than seeing this as any influence from Hinduism, it is much more likely to be a reflection of those elements in Zarathustra's teaching that have to do with the protection of animals — especially the Cow — from harm, as well as ancient Persian laws that codified abuse of certain animals, such as cruelty towards dogs, as a capital offense.

Pythagoras is known as the founder of mathematics in the Western world. At the core of Pythagoras' mathematics there is a binary opposition between *perâs* or "limit" and *apeiron* or the "unlimited." The world is brought into being through a dynamic interaction between the former principle of order and the negatively infinite or unlimited, in other words *chaos*. The place where we see this elaborated in the greatest detail is a text of Plato that the majority of scholars believe is the most Pythagorean book that he wrote. It bears mentioning that Plato was, of course, a member of the Pythagorean Order. Plato's "creation myth" in the *Timaeus* portrays chaotic unformed matter being shaped by mathematical principles of order, limit, and proportion, so that the good creation can come into being.³³ The most renowned idea of Plato, that of incorruptible perfect and eternal archetypes, the *eidos* or "forms" of all existing things — plants, animals, human beings — has no precedent whatsoever in Greek thought, but it most definitely does in Zoroastrianism. *Ahurâ Mazdâ* primordially creates a world of perfect archetypes, the *fravartis* or *fravashis*, of all natural beings in a luminous state (*eidos* suggests something that shines in its appearance) before they are assaulted from out of darkness by the chaotic and deranging forces of the Lie.³⁴

The deepest Persian influence on Plato, via the Pythagorean Order, was probably in the domain of his political philosophy. The Greeks had a number of traditional forms of political organization: *Democracy*, *Oligarchy* (rule of the wealthy), *Timocracy* (martial law), *Tyranny* (the arbitrary rule of one absolute dictator). Relatively unstable city states frequently alternated between these types of regime. For example, a democracy would wind up being taken advantage of by wealthy people who manipulated the ignorant masses to establish an oligarchy, and then perhaps some champion of the people would rise up from out of the military, initially promising to eliminate corruption together with his comrades, but eventually murdering his fellow generals and establishing himself as a tyrant. By contrast in ancient Iran, for centuries before the classical Greek period, there was a well-established tradition of the alliance between the philosopher and the king. This goes back to Zarathustra's own relationship with Kavi Vishtaspa (Kay Goshtasp).

Pythagoras adopts this system of sovereignty grounded on the reverence for Wisdom. When he founds his school in southern Italy, he trains the children of the feudal landowners with a view to restructuring the government there into a genuine *Aristokratiâ*, namely a meritocracy wherein the most intelligent and competent people are making policy on the basis of expert knowledge and under the guidance of a single

---

33   Plato, *Timaeus* (New York: Penguin Classics, 1977).

34   Seyyed Hossein Nasr, "Bundahisn" and "Greater Bundahisn" in *An Anthology of Philosophy in Persia* (New York: Oxford University Press, 1999), 6–22.

chairman who is essentially a philosopher-king.³⁵ This project met with ferocious resistance. Eventually there was a coup, wherein the feudal lords burned down the Pythagorean schools and Pythagoras either died in that fire or he barely escaped and died of his injuries shortly thereafter.³⁶ Plato, who inherits this idea of Guardianship of the Wise from the Pythagorean Order, also tries to implement a philosopher-kingship in Syracuse and is almost martyred as well. He is forced to leave the city in disguise by the cover of night, once custodians of customary Greek culture in the court of the young man he was trying to influence managed to regain control of the situation.

The strongest objection that one might level against a Zoroastrian influence on Plato actually winds up being more evidence in favor of it. This has to do with his concept of the Noble Lie. One must place this concept in its proper context within the pages of Plato's *Republic*. At one point he claims that lies are only acceptable as tactics in warfare or stratagems employed in combat against the enemy.³⁷ Yet we also see noble lies employed with the aim of reorganizing society, especially on the basis of a revision of fundamental religious beliefs. How can this be considered an instance of military strategy or an intelligence operation against an alien enemy?

Throughout the *Republic* there is a sustained critique of Homeric epic poetry and the kind of moral values that it is inculcating in the Greek youth. On a number of occasions, Plato comes very close to saying that the Homeric myths are lies and that they are corrupting the minds of the youth.³⁸ Since this was such a dangerous thing to say — after all his own teacher, Socrates, was executed under the mere suspicion of his having impiously rejected the Olympian pantheon in favor of other religious ideas — Plato has to be careful about how he says it. What he winds up claiming is that we cannot accept that God would shape shift in order to deceive people, the way that Zeus often assumes the forms of animals

---

35   Guthrie, *The Pythagorean Sourcebook*, 36–37.

36   Ibid., 134–135.

37   Plato, *The Republic of Plato* (Basic Books, 1991).

38   Ibid., 56–57, 59.

to rape women. These ought not to be accepted as true stories because otherwise they will corrupt the morals of the youth, and besides we know very little with certainty about such by-gone prehistoric epochs and so we should take these fables and spin them in a way that will be more constructive for the cultivation of virtue.[39]

At its most incisive this critique of customary Greek culture comes to the verge of claiming that there is some such God as Zeus but that he is a liar, a great deceiver. Perhaps Plato is using the Noble Lie as a stratagem or tactic in order to combat a Godfather who is a liar. From 382e–383a in *Republic*, Plato writes:

> Nay, no fool or madman is a friend of God… Then, there is no motive for God to deceive? None. So, from every point of view the divine and the divinity are free from falsehood. God is altogether simple and true in deed and word, and neither changes himself nor deceives others by visions or words, or the sending of signs in waking or in dreams. You concur then, as our second norm or canon for speech and poetry about the gods that neither are they wizards in shape shifting nor do they mislead us by falsehoods in word or deed.[40]

Well, the very messenger or courier of Zeus is Hermes the trickster, liar, and thief — who, by the way, is also responsible for the ritual slaughter of cattle that Zarathustra is so indignantly opposed to in his *Gathas*. What kind of a god is this who employs a career conman as his messenger to humanity? Plato knows that the most high god of the Greeks is a deceiver, essentially Zarathustra's *Angrâ Mainyu* (i.e. *Ahriman*), the promulgator of the Lie. He has seen this only because he is under Persian influence. This is a shattering, culturally catastrophic recognition. It may be that Plato advocates the strategic use of deception only in response to the horrifying need to wage a military conflict with a demonic being who is a master manipulator of benighted human societies.

The most revolutionary aspect of the political system that Plato intended for his philosopher kings to unfold in Greek society is the equality of women, or at least the equal opportunity of women to serve in all

---

39 Ibid., 59–60.
40 Plato, *Collected Dialogues* (Princeton: Princeton University Press, 1999), 630.

capacities in the society. This, again, is an idea that has no precedent in Greek culture. Some scholars have suggested that this is possibly a sign of Spartan influence, but women in Spartan society were only marginally more free than in Athens where they had no political or property rights and where the most accomplished and independent women were prostitutes. Spartan ladies had the kind of independence that frontiers women did in the old American West, because Sparta was a militaristic society wherein men were often at war so that women had to be relatively more resourceful in order to, as it were, man the homestead. If we really want to look at the society in Plato's time where women had a completely different status, then we have to look at Achaemenid Iran. Women in the Persian Empire were property holders. They had their own estates and people employed in their service on those estates.[41] If we look at the records that are signed with the personal seals of these ladies, and the letters that these women wrote, we can see that women were even paid at a rate equal to the salary of men for specialized labor.

Greek accounts are rather critical of Persian men for allowing their women to have tremendous influence over those of them who were in leadership positions. They mocked Persian men for being under the thumb of their women. We also have a few examples where Persian women were commanders or admirals in the military, whether in the Immortal Guard or the Persian Navy. This is consistent with what we read in the oldest of Zoroastrian scriptures: "Thy good dominion, Mindful Lord, may we attain for evermore: may a good ruler, whether man or woman, assume rule over us in body and mind, O beneficent of beings."[42] Zarathustra consistently refers to men *and women* when he demands that each individual conscientiously exercise his *or her* free will and his respect for the wishes of his daughter, Porouchista, at her wedding ceremony, is an especially colorful vignette in the *Gathas*.[43] Unlike in mainstream Hinduism, or in

---

41 Maria Brosius, *Women in Ancient Persia* (New York: Oxford University Press, 1998).
42 M.L. West, *The Hymns of Zarathustra* (New York: I.B. Taurus, 2015), 176.
43 Ali Jafarey, *The Gathas, Our Guide: The Thought-Provoking Divine Songs of Zarathustra* (Cypress, CA: Ushta Publications, 1989); Piloo Nanavutty, *The Gathas of Zarathustra: Hymns in Praise of Wisdom* (Ahmedabad: Mapin Publishing, 1999).

the original teachings of Buddha, according to which a women needs to be reborn as a man before attaining perfect enlightenment, let alone the utterly misogynistic Abrahamic religions, in Zarathustra's teaching women have the potential to be men's equals because each person is regarded first and foremost as an individual soul. This is Pythagoras and Plato's prototype for women receiving an equal education to men, including military training, so that they would be fit to serve in the highest leadership positions of a justly governed society. Even Plato's notion, absurd to his contemporary Greeks, that leading women should be offered maternity leave or some kind of childcare so that they can pursue a career, finds its only pre-modern precedent in ancient Persian society.[44]

Throughout the entire classical Greek and Roman periods, to say nothing of the dark age that followed Europe's Christianization, Iranian women were far more free to flourish spiritually and materially than their European counterparts. The closest pre-modern European point of comparison to the status of Iranian women would be to Viking culture, and this is no accident because the Goths rode together with an Iranian tribe known as the Alans and were deeply influenced by them. In particular, the Alans, who were a branch of the Sarmatians, brought the tradition of *Javânmardi* or Chivalry into Europe. Numerous symbols, rites, and rituals of Chivalric culture appear in Iranian Civilization long before they do in Europe. These include: horseback warfare with chainmail, armor, and lances;[45] the sword in the stone trial of strength; the mystery of the grail; the knights of the round table; and, most relevantly to our present subject matter, the idea of knights sworn to be the champions or *Pahlavâns* of land-owning noble women who were also the object of romantic love from roguish traveling poets and minstrels.

The woman who owns her own estate and who is free to engage in romantic love affairs rather than being condemned to be the property of her father or husband is, in the first instance, an Iranian woman who has traveled from the Scythian and Sarmatian realms of the Iranian Caucasus

---

44 Brosius, *Women in Ancient Persia*.
45 C. Scott Littleton and Linda A. Malcor, *From Scythia to Camelot* (New York: Routledge, 2000), 8.

and the Black Sea to Brittany, Occitan (southern France), and Catalonia. She is a member of that Sarmatian warrior society where women were so revered that the Greeks mythologized them as Amazons. Although the *Shâhnâmeh*, largely a legacy of these northern Iranian barbarians, is probably the greatest treasure trove of the Chivalric ethos, one also sees *Javânmardi* or Chivalry in the conduct of the ancient Persian Emperors.

So the Iranian co-constitution of European civilization does not end with the impact of the ancient Persians on the classical Greeks. Chivalry and romantic love originated in the Iranian cult of Mithra, the Zoroastrian world savior. Not only are most of the positive elements of European 'Christianity' Mithraic in origin, the same tradition catalyzed the Mahayana reform of Buddhism in Eastern Iran and could have unified the entire Indo-European world.

Zarathustra (Zoroaster) was not a Zoroastrian. He preaches against a god of rapine and plunder whose soma-drunken devotees decorated their dwellings with the horns of sacrificed bulls, a warlord that Zarathustra accuses of being a great deceiver. Since the proto-Vedic culture is the context of the Iranian visionary's revolutionary teaching, it is not a stretch to imagine that he was opposing the worship of Indra. Before Zarathustra Indo-European religion bears no trace of the division between the gods (*Daevâs*) and titans (*Ashurâs* in Sanskrit, *Ahurâs* in Persian). This distinction may have its origin in Zarathustra's moral inversion that demonized the *Daevas* (Persian *divs*, devils). As the Vedic religion differentiates itself from that of the Iranians, Mitra, who was as important to the earliest Aryans as Indra, eventually disappears from Hinduism.

Mithras was the god of contracts, oaths, truth and trust. The handshake, as a sign of trust (that one is not carrying a weapon) in greeting a friend or closing a business deal, originated within the Mithraic order.[46] There is no evidence for it as a common custom in Europe prior to the spread of Mithraism from Iran to Europe. The earliest depiction of this is a first century relief of Mithras from Kurdistan, shaking right hands

---

46 Payam Nabarz, *The Mysteries of Mithras* (Rochester, VT: Inner Traditions, 2005), 25.

with the Parthian Shah Antiochus.[47] The Mithraic handshake was referred to as *dexiosis* and so initiates of the Order were called *syndexioi*, "those who have been united by a handshake" (with the Father).[48] So did the military salute, which symbolizes shielding one's eyes from the brilliance of Mithras, the general and Invincible Sun.[49] Since Mitra was the god of wisdom and truth, in the sense of trust and oath-taking, could it be that Zarathustra's gospel of Ahura Mazda is a reformation of the Mitra cult — one that pits its adherents in direct opposition to the worshippers of Indra?

In the 10th Yasht of the *Avestâ*, Mitra (Sanskrit) or Mithra (Avestan, ancient Persian) is made the equal of Ahura Mazda, when the Lord of Wisdom says to Zarathustra: "Verily, when I created Mithra, the lord of the wide pastures, I created him as worthy of sacrifice, as worthy of prayer as myself, Ahura Mazda."[50] Together with his virgin mother and partner Anahita, Mithra becomes the most important of the *yazatâs* in the religion of Zarathustra. Since he was preaching worship of an abstract creative intelligence, difficult for any but philosophical minds to grasp, Zarathustra integrated various beneficent pagan deities into his gospel in the form of *yazatâs* (*Yazdân* is one word for a "deity" in Persian). Forming battle lines against the demonized *Daevâs*, these titanic gods and goddesses (*Ashurâs*) become helpers of Mazda. Among them Mithra is unique. He hypostatizes the archetype of the *Saoshyânt* or World Savior that we see in the *Gathas* of Zarathustra.

Mithra, or *Mehr* in contemporary Persian, means "Light", "Love", and "Friend." He was born of his virgin mother in the middle of the night from December 24th to 25th, which (by the reckoning of ancient calendars) is the Winter Solstice — the rebirth of light from out of the most encompassing darkness.[51] This is celebrated at *Yaldâ* (an Indo-European

---

47   Ibid., 26.
48   Ibid.
49   Ibid.
50   Ibid., 4.
51   Ibid., 6.

cognate of *Yule Day*), one of the four most sacred Zoroastrian holidays still commemorated in Iran. Mithras was the "lord of green pastures" and the evergreen tree represented Truth, evoking his status as the god of trustworthy Oaths and Contracts, so *Yalda* was celebrated by bringing an evergreen into an enclosure and giving it gifts. Unlike the contemporary practice of this Yule tree tradition, the evergreen — usually a Cypress (*Sarv*) in Iran — was brought in together with its roots and then re-planted after the holiday. Mithra wears a red Phrygian cap, evoked by the *Mitre* of the *Pater* (Persian *Pedar*, "Father" or *Pir* in Sufism), as well as a white belted red cloak and trousers — a distinctively Iranian garment of Parthian and Scythian riders that spread to Europe only later, in large part through Mithraism.[52] Are you reminded of Santa Claus? Devotees of Mithra celebrated holy communion, with wine and loaves of bread that were impressed with the symbol of an equilateral cross inside of a circle — a reference to the equinoxes and solstices of the Invincible Sun (*Mehré Jâvedân* in Persian, *Sol Invictus* in Latin).[53] Baptism was also practiced, since Anahita is the goddess of the holy waters and the Lady of the Lake. She virginally conceives of the avatar of Mithra by submerging in Lake Hamun (in Sistan), where legend has it that the seed of Zarathustra was preserved.

The similarities of Mithraism to Christianity frightened the Christian writers who became aware of them, and they resorted to claiming that the devil, who had the demonic power to attain foreknowledge of the coming of Christ, had imitated elements of what would become Christianity and introduced them into the world in order to denigrate them and to misguide people.[54] Obviously, Zarathustra would have seen it the other way around: the Great Deceiver distorting Mazda's World Savior. In *The Error of Pagan Religions* (350 AD), the early Christian evangelist Firmicus Maternus chastises his fellow Romans for their Persianization:

---

52 Ibid., 49.

53 Ibid., 27.

54 Ibid., 13.

The Persians and all the Magi who dwell in the confines of the Persian land give their preference to fire and think it ought to be ranked above all the other elements. So they divide fire into potencies, relating its nature to the potency of the two sexes, and attributing the substance of fire to the image of a man and the image of a woman. The woman they represent with triform countenance, and entwine her with snaky monsters. …The male they worship is a cattle rustler, and his cult they relate to the potency of fire, as his prophet handed down the lore to us, saying: *mysta booklopies, syndexie patros agauou* (initiate of cattle-rustling, companion by hand-clasp of an illustrious father). Him they call Mithra, and his cult they carry on in hidden caves. …Him whose crime you acknowledge you think to be a god. So you who declare it proper for the cult of the Magi to be carried on by the Persian rite in these cave temples, why do you praise only this among the Persian customs? If you think it worthy of the Roman name to serve the cults of the Persians, the laws of the Persians…[55]

The core symbols and rites of Mithraism that are appropriated by Christianity date from at least the third century BC, the start of the Parthian dynasty. The Parthian princes, who perfected the culture of knightly chivalry (*Pahlavâni, Javânmardi*) practiced by their Scythian and Sarmatian cousins in the North, were devotees of Mithra who engaged in a centuries-long struggle with the Roman Empire. Oddly enough, by the first century BC the Roman legionaries at war with Parthia began adopting the Iranian religion en masse and built Mithraeums everywhere they were stationed in Europe, from Palestine in the East to Spain in the southwest and Britain in the northeast.

The Mithraic legacy in Britain or Celtic Brittany and in southern Spain is particularly noteworthy because it did not only consist of indirect Iranian influence via the Roman legionaries. A group of Iranians from the Sarmatian confederation, whose warrior women and leading Ladies the Greeks mythologized as Amazons, rode from their homeland in the Iranian Caucasus all the way across Europe.[56] They are called *Alans* (*Alani* in Persian) and the name *Alan* in European languages derives from them. In fact, *Alani* is a linguistic corruption of *Arâni* (think of how the Japanese turn an *r* into an *l* when they pronounce it). *Arân* was the ancient Iranian

---

55 Ibid., 11.

56 Littleton, *From Scythia to Camelot*, 8–11.

name of the Caucasus region. It is a contraction of *Ar* — the Indo-European root of the Persian words *Arya* and *Ar*ta (Truth) and of Greek words such as *Ar*istocrat, *Ar*ete (Virtue) — and *an*, which is a place name designator, as in Abad*an* (well-built place), or in Gorg*an* (place of the wolves).

Speaking of Gorgan, its similarity to *Gorgon* is not incidental. The city is located in the Caspian region of Iran that was once home to the Sarmatians, whose "Amazon queen" Tomyris defeated Cyrus the Great. The Sarmatians worshipped Gorgons, terrifying amphibious goddesses entwined with serpents, and their standard or war banner was a dragon or feathered serpent. This, taken together with their head-to-toe lizard-like scales — the first known use of chainmail armor — accounts for their having been dubbed Sauromatians (*Saur* as in Dinosaur) or "the reptilians" by the Greeks.[57] In fact the Iranian name *Sarmat* comes from *Sarm* or *Salm* (note the transformation of *r* into *l* again), one of the three sons of the Shahanshah Fereidun. While his sons Iraj and Tur were respectively bestowed with the realms of central Iran (Persia) and Scythian Central Asia, Sarm inherited the cold (Persian *Sarmâ*) region to the northwest, from the Caucasus to the Carpathians. Before the arrival of the Turkic Huns, Hungary was Sarmatian and its language was consequently Indo-European rather than Finno-Ugric.[58] Later, the Alans would fight to defeat Attila the Hun and protect Europe from non-Aryans.[59]

In other words, the Sarmatians or Alans are the European Iranians — very closely related to the Scythians in language and culture. The Ossetians of Georgia still speak the Sarmatian language, which they call *Iron* (a cognate of *Iran*). From the time of Marcus Aurelius to the fifth century AD, successive waves of these Alans deeply penetrated the European continent. Before the fall of the Roman Empire, defeated Alans were recruited as elite guards to defend Rome against the barbarians — including as far afield as Hadrian's wall in Britain.[60] After the fall of

---

57 Ibid., 13.

58 Ibid., 16.

59 Ibid., 244.

60 Ibid., 18–21, 26, 262–263.

Rome, the Alans assumed leadership of the other "barbarians", including and especially the Goths, with whom they co-founded a kingdom in Spain that was consequently named *Goth-Alania* and has now come to be pronounced Catalonia. It is from here that the Troubador culture spread throughout the rest of Spain and French Occitan.

The culture of the Goths was so deeply impacted by the Alans that, from the Roman perspective, the Germanic Goths were Iranian Scythians.[61] Another source of confusion was the these Iranians had a 'Nordic' phenotype in common with the Germanic Indo-Europeans, looking more the Goths than the Goths did like the average Roman.[62] Having established themselves as landed feudal gentlemen by the last days of the Roman Empire, these Iranians formed much of the political and intellectual elite in medieval France, northern Italy, and Spain, with the Celtic and Germanic barbarians living under them as village peasantry and being employed by them as shock troops.[63] The Alans were the top advisors to the Merovingian kings, and the bloodline of the Carolingian dynasty of France directly descends from these Iranians.[64] Alan families were dominant in Gaul from their arrival in the 5th century until the 1200s.[65] This is the period during which, in this very region, the Grail romances were written.

The culture reflected in the *Nibelungenlied* and *Vulsunga Saga* is that of this period, where Germanic tribes were living under the influence of an Iranian elite in Western Europe.[66] The word *saga* itself is simply the name "Scythian" in the native Iranian language of the Scythian, Sarmatian, and Alanic clans, namely *Sagâ or Sakâ* — as the Persian word *sag* or "dog". These northern Iranians were the first people to have deliberately bred hunting dogs, another tradition that they passed on to the Germans. The

---

61 Ibid., 27.

62 Ibid., 132.

63 Ibid., 26–35.

64 Ibid., 244.

65 Ibid., 248–249.

66 Ibid., 272.

Persians referred to the Scythians as *Sagzi* or *Saxi* or "dog people", for example in the *Shâhnâmeh*, where Ferdowsi refers to the epic's greatest hero as *Rostamé Sagzi* or *Saxi*. By the way, you are right to hear in this *Saxson*, which means "Scythian son". To be "Saxon" means to be descended from the Iranians of Europe. Before being transplanted to Germany, the terms "Upper and Lower Saxony" referred, in Persian, to the Scythians of the northwest, namely the Sarmatians, and those of the southeast, namely the Indo-Scythians of Sistan and Baluchistan whose most famous hero was Rostam. He was among the Scythians who migrated down into an area stretching from the Iranian province of Sistan through present-day Afghanistan and what we — anachronistically — think of as northern India.

Again, this makes the point that the civilizational boundary between Europeans and the eastern Aryans is an artificial construct that reflects the later conflict between Islam and the Crusaders. The distinction is not only anachronistic, it is also ironic, since the Iranians who brought chainmail, armor, lances, and the entire culture of knightly chivalry into Europe were, together with their Germanic cousins, among the foremost Crusader knights. The so-called "Germanization of Christianity" would be more accurately described as an Alanization of Christianity, since Alans formed the clerical elite of Europe as this took shape.[67] Of course, this Aryanization of a non-Aryan religion would not have been necessary had Constantine and his colleagues refrained by committing civilizational treason in the promotion of Christianity, rather than Mithraism, as the dominant religion of Rome.

Nevertheless, Mithraic chivalry and Grail mysticism survived in Europe. An exhaustive study by C. Scott Littleton and Linda A. Malcor, entitled *From Scythia to Camelot*, demonstrates that all of core values, idea, symbols, myths, and rituals of the Chivalric culture of Medieval Europe are Iranian, rather than Celtic or Germanic in origin.[68] For more than a century, scholars have described tales like Wolfram von

---

67 Ibid., 245.

68 Ibid., 281–283.

Eschenbach's *Parzival* as "Germano-Iranian" in origin purely on the basis of comparison to Persian literature.[69] Littleton and Malcor show how the intermediary between Persian chivalry and romantic literature and that of Medieval Europe is the Sarmatian culture, identifying the original sources of the legend of Lancelot,[70] the Lady of the Lake,[71] the Sword in the Stone,[72] Serpentine and Dragon symbolism,[73] the Kingly glory of the Holy Grail,[74] and the Once and Future King.[75]

All of these have their origins in the Sarmatian *Nart* (i.e. *Nord*) Sagas that are still preserved among the Ossetians, the Iranians of Georgia. Their primary deity is a goddess of Wisdom called *Shatânâ*, symbolized by a serpent or dragon, who emerges from out of a fairy land or otherworld of Amazons deep beneath lakes, in order to choose and train the knightly hero haloed by the royal glory.[76] This figure, from whose name the Zoroastrian Persians derived *Shaytân*, which was then masculinized and imported by the Jews and Arabs into their pantheon as the demonized "Satan", is also archetypally identical to Artemis or "the Scythian Diana", as well as the figure of Anahita in Mithraism.[77] *Shatânâ* stands to the chosen chivalric hero-savior among the Knights of the Round Table as Anahita does to the savior-hero, Mithra. While we are on the subject of the Knight of the Round Table, it is worth noting that the idea has its roots in the free-spirited and meritocratic egalitarianism of Mithraism, wherein each nobleman saw himself as a king so that their leader was simply a King of Kings, an acting chairman seated at a round table of knights.

---

69 Ibid., 213–215.
70 Ibid., 79–124.
71 Ibid., 153–179.
72 Ibid., 181–194.
73 Ibid., 195–208.
74 Ibid., 216, 237–240.
75 Ibid., 70.
76 Ibid., 133, 153–162.
77 Ibid., 164–165.

That the Chivalric ethos of Germanic Europe is Iranian in origin is also supported by the fact that one of its masterworks, *Tristan and Isolde* is modeled on *Vis and Ramin* — an epic romance originally dating from the Parthian period (247 BC–224 AD), the era wherein the Mithra cult was most dominant in Iran.[78] No earlier example of a convention-defying romantic love, and the roguish rescue of a lady in waiting, can be found than this tale which, after the Islamic conquest of Iran, was preserved by the 11th century Persian poet Fakhruddin Gorgani.

Aside from its influence on specific literary works or elements of Grail mysticism, the Parthian culture of Iran, the culture of Mithra's princes, has a thoroughly medieval European knightly aesthetic — centuries before 'Gothic' culture in Europe. When one looks at the architecture, statuary, and stone reliefs at Parthian archeological sites, or at crafts products from the Sassanian period that preserve Parthian themes, and compares these to the contemporaneous aesthetic of the pagan Roman Empire, then to the atmosphere of medieval 'Christian' Europe, the direction of influence is quite clear.[79]

But how could Mithraism have become the dominant religion of the pagan Roman Empire, even though it was the cult of Rome's rival superpower? The Parthian king Mithradates (195–132 BC), whose name means "Mithra's Justice" (*Mehrdâd*) in Persian, had either set up or bought out the Cilician pirates of the Mediterranean Sea and was employing them essentially as a black ops or false-flag Persian Navy. Ostensibly stateless pirates driven by private avarice, they actually answered to Mithradates and were the vanguard of Mithraism in Europe. These Mithraists, to whom Anahita as a goddess of the waters was also no doubt very important, had connections to aristocratic houses all across the Roman Empire.[80] Eventually they grew so formidable that the Roman Navy did not have

---

78 Dick Davis, "Introduction" in Fakhraddin Gorgani, *Vis and Ramin* (New York: Penguin Classics, 2008).

79 Ghirshman, *Persian Art: Parthian and Sassanian Dynasties, 249 B.C. — A.D. 651.*

80 David Ulansey, *The Origins of the Mithraic Mysteries* (New York: Oxford University Press, 1989), 88–90.

freedom of movement in the Mediterranean. Even Julius Caesar was once taken prisoner by them on the high seas.

The cult of Mithras was well suited to Pirates on account of its extraordinary egalitarianism, itself an outgrowth of Zarathustra's teachings against the Vedic caste system and his emphasis on individual conscience. In the secrecy of a Mithraeum, merchants were on an equal footing with upstanding citizens, who treated slaves as their equals and in this occult order even a common soldier who had attained the highest rank of initiation could be looked up to by an emperor.[81] We can see this from the fact that just after Constantine adopted Christianity as the official state religion, a failed Mithraic restoration of sorts was staged by Caesar Julian (336 AD) who was himself a devotee of Mithras.[82] Such was the threat to Roman national security by the rise of Mithraism as the de-facto religion of the Empire that, Roman elites around Constantine may have seen the adoption of Christianity as a bulwark to guard against a potential military coup that would have erased the border between Iran and Europe — as if it would have been a bad thing to reunify the Indo-European world!

A reunification of the Indo-European peoples through the cult of Mithra (i.e. Mithras/Mitra/Maitreya) would in all likelihood have even extended into those parts of Asia that had embraced Mahayana Buddhism. An Iranian reform of the Buddha Dharma, Mahayana took shape under Zoroastrian influence. The Mahayana doctrine was developed by the Kushan dynasty, Iranians of the Scythian tribe who swept down into northern India from Khorasan. They are cousins of the free-spirited riders who established the Parthian dynasty in Western Iran. North Indians referred to the Kushans by one of the same names used by the Parthians, *Pahlavân* — meaning "champion" and the word from which we get *Pahlavi* or the name of the Middle Persian language. Since the Parthians clearly emerged from a Mithraic culture in Khorasan, which we see reflected in the chivalrous romances and tragedies preserved by Ferdowsi in the *Shâhnâmeh*, it is reasonable to assume that the Kushans, who are known

---

81 Nabarz, *The Mysteries of Mithras*, 12.

82 Ibid., 18.

to have been some kind of Zoroastrians, also revered Mithra. The central figure of Mahayana is *Maitreyâ*, the Buddha To Come, a savior figure that is anathema to the Buddha Dharma in its orthodox form as preached by Gautama and whose name is clearly a variant of Mithra.[83]

Mâhâyânâ or "the Greater Vehicle" emerged when the Kushan King Kanishka the Great (127–163 AD) convened the largest Buddhist council in history at Kashmir, led by his advisor Ashvaghosha. Under the influence of the Iranian king the monks and scholars worked out a new vision of the Dharma.[84] The founding texts of the new doctrine were engraved on copper plates, not in Sanskrit but in Kharoshti, a Hellenistic script designed by Kanishka to express the Iranian language of the Kushans. Fundamental changes in doctrine shepherded by these men reflect Iranian influence.

The first and foremost of these is the distinction between an esoteric teaching and an exoteric teaching. This was also a key feature of Western Iranian and Roman Mithraism. It is the means by which the Mahayanists could retro-actively claim that Gautama did teach their "Greater Vehicle" to initiates and that the *Therâvâdâ* doctrine that is his only publicly recorded teaching (in the *Tripitâkâ*) was a *Hinayânâ* or "Lesser Vehicle" calibrated to those with lesser understanding. The great difference between the two is that the Lesser Vehicle (which impartial scholars know to be Gautama's actual teaching) focuses on individual escape from the cycle of karma in a world of misery. In its view a Buddha arises only once over vast cycles of time, so that the idea one could become a Buddha in one's own lifetime is absurd. Whereas the Greater Vehicle holds that it is possible to attain perfect enlightenment here and now; the whole of creation is headed towards perfection, a Utopian society where all individuals will be delivered from suffering — essentially Zarathustra's *frashgard* (*frashokereti*).

Instead of life being seen as suffering or perpetual dissatisfaction (*dukkhâ*), with the goal of life being extinction (*nirvana*), life comes to be seen as a blessed opportunity for self-perfection. In place of Gautama's

---

83 R.E. Emmerick, "Buddhism Among Iranian Peoples" in *The Cambridge History of Iran, Volume 3(2)* (New York: Cambridge University Press, 1996), 949–964.

84 Richard Frye, *The Heritage of Persia* (Costa Mesa: Mazda Publishers, 1993), 223–224; Emmerick, "Buddhism Among Iranian Peoples," 995.

teaching that the ultimate nature of reality is nothingness (*shunyâtâ*) and that one's selfhood is an illusion to be overcome (*anâttâ*), Mahayanists put Zarathustra's teaching about archetypes of our perfected light bodies into the mouth of their Buddha. Music, dance, art, and eros are condemned by Siddhartha as conduits of destructive desire and as behaviors that trivialize life's all-pervasive misery.[85] Prior to the Kushans there is hardly any 'Buddhist art.' Under Kanishka's guidance we see merry making and even ecstatically erotic practices become sacraments of a Mahayana tradition that also generates a life-affirming 'Buddhist' art and architecture in a syncretic Greco-Persian style.[86] Shakyamuni claims that women need to be reborn as men before attaining enlightenment whereas, like Zarathustra and the ancient Persian Emperors, Mahayana recognizes that women are the spiritual equals of men and are in some cases even fit to be their teachers.[87] This was a view well suited to the horseback Iranian tribeswomen that Greeks mythologized as Amazons.

Under the influence of Zarathustra's concept of *Shahrivar* or "Desirable Dominion", unequivocal pacifism was rejected as a sign of Enlightenment. The *Dharmâ Râjâ* is no longer viewed as a beneficent but still unenlightened ruler, who would reject violence altogether if he were to attain Buddhahood. Instead, a perfectly enlightened sage can also be a ferocious warrior. Bodhidharma, a blue-eyed and red bearded Iranian, even developed a Buddhist martial arts tradition that he carried into Asia with him as part of the Chan or Zen teaching that was his particular contribution to Mahayana.[88] Padmasambhava, another ethnic Iranian from Khorasan,

---

85 Bhikkhu Bodhi, *In the Buddha's Words: An Anthology of Discourses from the Pali Canon* (Somerville, MA: Wisdom Publications, 2005), 175, 245.

86 Heinrich Zimmer, *The Art of Indian Asia: Its Mythology and Transformations* (Princeton: Princeton University Press, 1968).

87 Keith Dowman *Sky Dancer: The Secret Life and Songs of the Lady Yeshe Tsogyel* (Ithaca, NY: Snow Lion Publications, 1996); Jérôme Edou, *Machig Labdrön and the Foundations of Chöd* (Ithaca, NY: Snow Lion Publications, 1996).

88 John J. Jørgensen, "Bodhidharma" in *Inventing Hui-neng, the Sixth Patriarch* (Leiden: Brill Academic Publishers, 2005), 110.

later epitomized the Mahayana archetype of the fearsome warrior who is nonetheless a perfectly enlightened Buddha.

The influence of the chivalric or heroic (*Pahlavâni, Javânmardi*) militarism characteristic of Mithraism is evident in these Iranian missionaries who carried Buddhism into Tibet and along the Silk Route further east to China and Japan. The first person to tread this path was the Persian prophet Mani, who was of Parthian lineage. Although he was not a Mithraist, that Mani explicitly tried to synthesize Zoroastrianism and Platonism with Buddhism, and even came to be known as "the Buddha of Light" as far east as China, speaks to the strategic depth and globalizing potential of Iranian influence during the period when Parthian Mithraism was the dominant religion of Iran and the rising de-facto faith of Iran's greatest rival, the Roman Empire.

In fact, *most* of the Buddhist missionaries who turned the religion into an East Asian tradition were of Iranian origin and hailed from Khorasan, which was then the epicenter of Mahayana. The greatest Buddhist temple ever built, *Now Bahâr* ("New Spring" in Persian) with its 93-meter golden dome, and the most colossal Buddha statues ever carved (at Bamiyan) were in eastern Iran. The Aryan bone structure, fair skin, blue or green eyes, and auburn hair of the missionaries who set out from here on journeys into Asia are clearly depicted in the beautiful cave murals that they left behind in places such as the Tien-Shan mountains of the western Gobi.[89] Some of these caves have been sealed off from the public by the Chinese government. The reason is that the scenes depicted in them are so erotic that they would apparently offend contemporary public morality as badly as they would have offended the morality of early Indian Buddhists. What we have here is the iconography of the Left Hand Path, and I would argue that the true origins of Tantra are Mithraic and Iranian.

---

89 William Watson, "Iran and China" in *The Cambridge History of Iran: Volume 3(1), The Seleucid, Parthian and Sasanian Periods* (New York: Cambridge University Press, 2003); Roderick Whitfield, Susan Whitfield, and Neville Agnew *Cave Temples of Mogao: Art and History on the Silk Road* (Los Angeles: The J. Paul Getty Trust, 2000); Ghirshman, Roman *Persian Art: Parthian and Sassanian Dynasties, 249 B.C.–A.D. 651* (New York: Golden Press, 1962).

The Germanic barbarians, who would later dominate the Romans, first took on their distinct ethnic and linguistic identity when they broke off from the common ancestors that they shared with the Scythians living in the region north and west of the Black Sea (present-day Ukraine, Moldova, and Bulgaria). Unlike the Persians, who adapted Semitic writing systems (Cuneiform, Aramaic) for the administration of their Empire, the Scythians had a runic writing system for their northern Iranian language that is similar to but older than the Germanic runes. One of these runes is particularly worthy of note. To this day the word *Tyr* means "arrow" in Persian. In Iran's mythology it refers to the arrow of Arash, the heroic bowman. The Persians were famous for archery, and this greatest of all archers lets loose a magical arrow that defines the scope of the rightful realm of Aryans (*Irân*) in distinction from the non-Aryan (*Anirân*) world.

In Greek, Indo-European *sh* syllables are often softened into *s* (as in *Kourosh* becoming *Cyrus*). So Arash becomes Ares, the god of war referred to by the Romans as Mars. Roman Mithraists conflated Mithras with Mars, a fact that provides some further context for how the cult of Mithras becomes the dominant religion of the Roman military. Now, in Nordic mythology Tyr is seen to be the equivalent of Mars or Ares. His rune symbol is the arrow, the one loosed by the bow of Arash. Finally, the Norse see Tyr as the god of oaths and contracts, exactly the same function that Mithra has in the religion of Zarathustra. The dog was considered among the most sacred animals by Zoroastrians, and the Persians who used dogs to guard their homes from demons, and considered cruelty towards dogs a capital offense, originated the tradition of the dog as "man's best friend."[90] Fenrir, the wolf that Tyr is trying to domesticate (or turn into a dog) bites off his right arm.

Instead of being seen negatively, the extended right arm of Tyr being bitten by Fenrir is the Norse symbol of swearing an oath — in other words a symbol of the Truthfulness and Trust that are at the core of Zarathustra's teaching and that Mithra in particular embodies. This is why Mithra is depicted, together with his mother Anahita, in Sassanian reliefs portraying

---

90 Arthur Gobineau, *The World of the Persians* (Genève: Editions Minerva), 24.

the investiture of Persian emperors (their oath of office). At the same time, the wolf leaves Tyr with only one hand to use — his Left Hand. Since it is easy to see how Tyr and Arash or Ares the bowman are a single figure, this means that the Aryan archer — our god of war — is forced to use his Left Hand. I believe this lies at the origin of the designation "Left Hand Path". That would also explain why the north Indian branch of the primordial Indo-European tradition warns that the Left Hand Path is only suitable for warriors or individuals with heroic natures (*Virâ*).

Ahurâ Mazdâ is the *ashurâ* or Titan of Wisdom, and Mithra is his great champion and world savior in the battle against the gods or *daevâs*. If Zarathustra's revolutionary teaching is what originally divided the Indo-European community, then I propose that is because it is the original form of the Left Hand Path or the Titanic (*Ahurâi*) Religion. It is also worthy of note that, when the Tantric core of Mahayana is ultimately distilled by Padmasambhava (who I remind you, was from Khorasan), it adopted the designation *Vajrâyânâ* or "Thunderbolt vehicle" suggesting the adamantine strength of the titan who steals Indra's scepter and makes it his own. The lightning strike or thunderbolt is, of course, another connection between the Norse tradition of Tyr or Thor and this titanic current of Buddhism that followed the silk route to Asia from Eastern Iran.

The glory of Mughal India (including present-day Pakistan) must also be attributed to the Aryan civilization of Iran, and not to Islam. Akbar the Great and his successor, Dara Shokouh, were Persianate rulers who tried to move beyond the conflict between Islam and Hinduism on the basis of ideas from the school of Persian mysticism with the most direct connection to the Zoroastrian tradition, the *Eshrâqi* or "Orientalist" school of Shahab al-Din Suhrawardi — a martyred medieval Iranian apostate.[91] The Mughal culture of Northern India was in, nearly every significant respect, an Iranian culture. Its administrative and literary language was Persian.[92] Its art and architecture were also predominately influenced by

---

91 Makhanlal Roychoudhury Sastri, *The Din-i-Ilahi or the Religion of Akbar* (Nabu Press, 2014).

92 Muhammad Abdul Ghani, *History of Persian Language and Literature at the Mughal Court: With a Brief Survey of the Growth of Urdu Language* (Gregg Publishing, 1972).

Iranian models.[93] For example, a Persian, Ustad Ahmad Lahauri, was the architect responsible for the *Tâj Mahal* (a name meaning "Crown District" in Persian). Akbar and his short-lived successor instituted a project of translating all of the major works of Indian literature into Persian.[94] These fifty or so books included the *Mâhabhârâtâ*, the *Râmâyânâ*, the *Vishnu Purânâ*, and the *Bhâgavâtâ Purânâ*. Until well into the period of British colonial rule, centuries after Akbar's time, the *Bhagavâd Gitâ* was more widely read in Persian in Northern India than in Sanskrit. It was said that this most revered of all Indian spiritual texts resounded better in Dara Shokouh's Persian translation, under the title *Âbé Zendegi* or "The Water of Life."

Most of northern India—well beyond Pakistan—has spent most of its history as part of Iranian Civilization. In addition to the Mughals, who established a Persianate culture, Persians ruled directly over northern India during the Achaemenid, Sassanid, and Afsharid dynasties, and as we have seen the Kushans—the promulgators of north Indian Buddhism—were also ethno-linguistically Iranian. In other words, Iran or *Irân-Shahr*—literally the "Aryan Imperium"—is the quintessentially Indo-European Civilization.

---

93 Catherine B. Asher, *Architecture of Mughal India* (New York: Cambridge University Press, 1992).

94 Carl W. Ernst, "Muslim Studies of Hinduism? A Reconsideration of Arabic and Persian Translations from Indian Languages" in *Iranian Studies*, Vol. 36, No. 2 (Jun., 2003), 173–195.

CHAPTER 7

The Indo-European World Order

For a public mind enslaved to mass media, the horrific nightmare of the 37 year long Islamic regime in Tehran has all but obliterated the great regard that Western intellectuals and artists had for Iran during most of the modern period. Goethe believed that Hafez was the greatest poet of all time, and he conceived of his *East-West Divan* on the model of the *Divan* of Hafez and as something of a reply to it.[1] G.W.F. Hegel's extensive writings on Iran in his *Aesthetics*, his *Philosophy of History*, and *Phenomenology of Spirit* have established his reputation among Iranologists as one of the founders of Iranian Studies.[2] It appears that Hegel's conception of the dialectical progression of History towards the society of Spirit is influenced by centuries old Iranian conceptions of cosmic evolution. Fakhruddin Gorgani's mystical romantic epic *Vis and Ramin* became the basis for *Tristan and Isolde*, the most sophisticated elaboration of which is the opera by Richard Wagner.[3] As we have seen, Friedrich Nietzsche read the first translation of the hymns of Zarathustra and adopted the Iranian prophet as his own literary persona in his masterwork, *Thus Spoke Zarathustra*, which was adapted into an orchestral piece by Richard Strauss. This was

---

1   Shafiq Shamel, *Goethe and Hafez: Poetry and History in the* West-östlicher Divan (New York: Peter Lang International Academic Publishers, 2013).

2   Henry Corbin, *The Voyage and the Messenger: Iran and Philosophy* (Berkeley: North Atlantic Books, 1998), 60–64.

3   Dick Davis, "Introduction" in Fakhraddin Gorgani, *Vis and Ramin* (New York: Penguin Classics, 2008).

not, however, the first instance of Zoroaster in Western music. Mozart's *Magic Flute* preceded it, with a dramatized version of *Sarastro* at its core. Edward Fitzgerald's much embroidered 'translation' of *The Rubaiyat of Omar Khayyam* became a major influence on Victorian English poetry. Emerson wrote a reverent essay on "The Persian Poets" which ought to make us wonder how deeply his transcendentalism was influenced by that of the Iranian texts that he devoured.

No account of Iran's impact on the modern West would be complete without a special emphasis on the French. Voltaire and Montesquieu were both quite interested in Iranian culture and their writings reflect it. Montesquieu satirized the European culture of the Enlightenment from an Iranian perspective in his *Persian Letters*, reaching back to the classical Greek literary practice of internal cultural critique by means of using the Persians as a foil — they reassume the position of that civilized "other" cultured enough to give Europeans some perspective on themselves.[4] Voltaire saw in Zarathustra a proto-Deist and thus a holy father more adequate to the spirit of the Age of Enlightenment than Jesus Christ.[5] Henry Corbin, the first translator of Martin Heidegger's writings into French — who, consequently, helped conceive of Continental Philosophy, which is largely a French reception of Heidegger — was the most prominent Iranologist of the 20th century. He saw in Shahab al-Din Suhrawardi and other thinkers in the still living School of Illumination (*Eshrâqiyoun*) visionary predecessors of Heideggerian hermeneutics.[6] Corbin came to believe that Iran would be "where the final liturgy setting the world on fire will take place."[7] It is beginning to look like he was right.

---

4 Charles-Louis de Secondat Montesquieu, *Persian Letters* (New York: Penguin Classics, 1984).

5 Francois Voltaire, "Zoroaster" in *Philosophical Dictionary* (New York: Penguin Classics, 1984).

6 Henry Corbin, "From Heidegger to Suhrawardi", an interview with Philip Nemo (recorded for Radio France-Culture on June 2, 1976).

7 Henry Corbin, *Spiritual Body and Celestial Earth* (Princeton: Princeton University Press, 1977), 50.

Since at least 2012 a grassroots movement has given birth to what the Pahlavi regime tried but failed to accomplish from the top down: a cultural revolution that restores and revitalizes the Pre-Islamic identity of Iran — an Iranian Renaissance. The term "Renaissance" calls to mind Medici Italy. It is more than an analogy. One thing that is clear to partisans of the Iranian Renaissance, and that Westerners are going to have to understand soon, is that the idea of "Arabic Science" or an "Islamic Golden Age" being a bridge between classical antiquity and the European Renaissance is grotesque nonsense. Repeating such absurdities adds insult to Islam's injury of Iran. When the Caesar Justinian of the Christianized Eastern Roman Empire of Byzantium closed the last of the great classical academies, the one at Athens, in 529 AD on the grounds of sacrilege, no less than seven of its Neo-Platonist masters resumed their careers at the Academy of Gondeshapur.

One of three major universities in Sassanian Iran, like the Library of Alexandria, the Academy of Gondeshapur featured laboratories for practical research as well as a hospital that was the most renowned medical facility in the world. Its name means "Great Shapur" since it was founded by Shapur I, the Persian Emperor whose vision of Zoroastrianism was so broadmindedly true to the "progressive mentality" of the *Gathas* that he invited Mani to his coronation ceremony to sermonize about how Zarathustra's esoteric teaching is one with that of Buddha and the Gnostic Christ. This is one of a number of examples of Persian Emperors aspiring to form an alliance with the leading visionary thinkers of their epoch, so that the two together could govern in the sagacious manner that Zarathustra and Goshtasp once did. This was the model for Plato's philosophical Guardians of the state. Indeed, when the last of Europe's classical academicians took refuge in Iran they dubbed Khosrow I (*Anushiruwân*) the ideal Platonic philosopher king.

The Sassanian interest in Neo-Platonism was such that extensive translations of Greek texts into Syriac, an administrative language of the Western parts of the Persian Empire, were already under way at the Persian ruled city of Nisibis (in present-day Syria). Once the academicians arrived in the Sassanian capital of Ctesiphon and were set up at Gondeshapur, these translation efforts were stepped up and these inspired original scientific treatises in Pahlavi (middle Persian) written by Bozorgmehr (Borzuya) and others.[8] Such scientific efforts were not only taking place under Greek inspiration, but were also catalyzed by extensive translations of Sanskrit Indian texts into Pahlavi. In other words, about a century before the Arab Muslim conquest of Iran began, there was an enlightenment underway that promised what (somewhat anachronistically) we might see as a Zoroastrian fusion of Western science with Eastern spirituality.

When the Arabs invaded in 651 AD they spent the first hundred years of their conquest setting fire to libraries and ransacking universities, including the Academy of Gondeshapur. What the Islamic State recently did in Mosul, where it burned every book in every library that was other than a canonical Islamic text, was just taking a page out of the playbook of the original Mohammedan armies that conquered Zoroastrian Iran. Only by 765 did the Caliphate begin to recognize that one cannot govern an empire this way. They pieced together what fragments survived their wonton destruction, aggregating the remaining texts at the House of Wisdom (*Bayt al-Hikmat*) in Baghdad (*Bogh Dâd*, which means "God's Justice" in Persian). The library of the new school was placed in the charge of a Persian, Abu Sahl al-Nawbakhti, who oversaw the project of further translating the Syriac and Pahlavi translations of Greek originals into Arabic, and in rarer cases translating Greek directly into Arabic. By this means the whole Aristotelian corpus and most of Plato were preserved in Arabic translations as they met their demise at the hands of zealous Christians in Europe.

The vast majority of the scholars of the so-called "Islamic Golden Age" were Persians, especially when mere translation gave way to commentaries and original texts inspired by Greek rationalism and science (which were probably catalyzed by the Zoroastrian colonization of Greece in the first place). These Persian geniuses of the medieval period were nearly

---

8   A. Tafazzoli and A.L. Khromov, "Sasanian Iran: Intellectual Life" in *History of Civilizations of Central Asia: Volume III The Crossroads of Civilizations A.D. 250 to 750* (United Nations Educational, Scientific and Cultural Organization: Paris, 1999), 91–93.

all from that part of eastern Iran known as Khorasan (extending into present-day Afghanistan, Tajikistan, and Uzbekistan), which remained demographically white or ethnically Aryan for centuries after the Arab conquest (and before the Mongol invasion). In other words, these people were like ethnic Germans (as in German Science) whose own fathers and grandfathers were still practicing Zoroastrianism and resisting the country's forcible conversion. Some of them were even referred to as "majusi" by Arab rulers of the time, an epithet for 'pagans' that derives from *magus* (a Zoroastrian priest). Together with the rebel stronghold of Azerbaijan and the Caspian coast, Khorasan was the site of the largest number of revolts against the Caliphate on the part of Persian fiefdoms trying to carve out some degree of autonomy.

These Persians are the ones who crafted a philosophical and scientifically adequate Arabic vocabulary, radically transforming the language of the desert tribesmen in order to translate complex Greek and Persian terminology.[9] It is a historical travesty of the first order to credit Islam with the brilliance of Razi (Rhazes), Ibn Sina (Avicenna), Khwarazmi (Algorithmi), Farabi (Alfarabius), Al Biruni, and Omar Khayyam. Every single one of these physicists, physicians, chemists, mathematicians, and astronomers was a Persian forced to write in a language other than his own because Iran was under Arab occupation. They are part of the treasure trove of the Indo-European or Aryan genius. The truth is that the "Indo" of the Indo-European world has always referred, not to Dravidian India at large, but to the southeastern border of at least six Iranian or Persianate kingdoms that extended to the Indus river for most of history. With its northwestern border in the Scythian realm from the European coast of the Black Sea to the Caucasus, and its southwestern border at the Indus, Greater Iran was *the* Indo-European civilization.

Today, after a generation of life under an Islamic theocracy more orthodox than anything Iran has suffered since the Arab Conquest, and with 70% of its population under the age of 30, the spirit of Zarathustra is

---

9 Victor Danner, "Arabic Literature in Iran" in *The Cambridge History of Iran: Volume 4, From the Arab Invasion to the Saljuqs* (Cambridge University Press: New York, 1999), 566–594.

returning with a vengeance. Conversion out of Islam is officially punishable by death. For this and other obvious reasons objective polls are impossible to conduct, but judging from a variety of fairly clear sociological markers — like how many young men and women wear *Farvahar* pendants, or how common the discourse of "Aryan" identity has become — something like one-fifth of Iran's population has left Islam and now identifies with the Pre-Islamic Persian ethos. Combine this with the fact that the Kurds, historically the most significant Iranians besides the Persians, are also returning to some form of the religion of Zarathustra (including forms in which Mithra plays a prominent part). In that region of *Irân-Shahr* commonly known as Kurdistan (and extending across three present-day nations states besides Iran proper), Neo-Zoroastrianism has become the most virulent reaction against the rise of the Islamic State — with its genocide of the Yazidi Mithraists and its destruction of Pre-Islamic Iranian archeological sites in areas that were part of three successive Persian Empires.

In Iran this Neo-Zoroastrian movement began after the failure of the Islamic reform movement, which culminated in the protests of 1999, and accelerated its pace following the brutal regime crackdown on the much larger uprising exactly a decade later in the summer and fall of 2009. The Aryan identity of Iran that binds her destiny with that of Europa is not a cultural-historical curiosity. It is the basis of the Iranian Renaissance, a cultural revolution triggered by the failed uprising against the Islamic Republic in 2009. If one were to form a projection on the basis of the current trend, should present social and political conditions persist in Iran for only another decade, the country is headed for a violent cultural revolution wherein a militant minority of about 30% of the population that has left Muhammad and Ali for Zarathustra and Mithra finally outnumbers the 15% who are pro-regime dead-enders. At that point, something awesome and terrifying will take place: a revaluation of all values and a polar inversion of the world (dis)order of the UN International System, established by the Allied Powers between 1945 and 1948.

As the Islamic State and Al-Qaeda (backed by Turkey, the Saudis, and the Arab Emirates) surround Iran in territories that have been part of one or another Persian Empire for most of their history, the ultimate consequence is clear. Once the Iranian or Aryan Renaissance triumphs domestically, the Persians and Kurds in the vanguard of the battle against the nascent global Caliphate — with its fifth-column in the ghettos of major European cities — will reconstitute Greater Iran as a citadel of Indo-European ideals at the heart of what is now the so-called 'Islamic world.' Whether contingent circumstances mean that it will be ten, fifteen, or even twenty years from now — this is *going* to happen.

Looking back at Samuel Huntington's argument in *The Clash of Civilizations* in light of the imminent, apocalyptic confrontation with Islam in Greater Iran, reveals a major flaw in his thinking. Huntington quotes Oswald Spengler to the effect that a civilization is the "destiny" of a culture in which technical development is taking place. There is something about the way that technoscientific thinking grasps the world, which will over the long-run transform all diverse traditional cultures in a single direction. By contrast, to equate civilization with culture, to say that a civilization is merely "a culture writ large" is to elevate cultural relativism to a civilizational level where it becomes an inevitable "clash of civilizations." This is nowhere more apparent than in Huntington's shocking statement that: "Civilizations are the ultimate human tribes, and the clash of civilizations is tribal conflict on a global scale."[10]

An Italian village may have a different culture than a German one, which in turn may be distinct from a French city, but the people of all three are Europeans, and while there may be some cultural differences between Europeans as a whole and North Americans, Canadians and citizens of the United States would join Europeans in identifying themselves as "Westerners." According to Huntington there is, however, no comparable umbrella of broad identification that would encompass Westerners, Chinese, Hindus, and Muslims. These peoples belong to distinct civilizations. Interestingly, although he does not take racial distinctions to be isomorphic to those of civilization (which would be manifestly false), Huntington makes the remark that: "A civilization is thus the highest

---

10 Huntington, *The Clash of Civilizations*, 207.

cultural grouping of people and the broadest level of cultural identity people have *short of that which distinguishes humans from other species.*"[11]

There are two huge assumptions here. First, that there is no conceivable "human culture" in general. Second, that if non-human beings were to have something like a culture or civilization it would necessarily be so alien that all human civilizations would be closer to one another than they are to these non-human forms of life. Not only is the second assumption highly questionable, it also contradicts the first.

Suppose technical development promotes evolutionary convergence beyond narrow-minded primitive cultures that may acquire and employ technology but without the first-hand philosophical experience of the revolutionary scientific discoveries that led to it. In that case, a non homo-sapiens culture from elsewhere might indeed have more in common with a certain human civilization than that civilization has in common with barbarian cultures on Earth.

Even though Huntington claims that "values, beliefs, institutions, and social structures" are what counts with respect to civilization(s), and not "physical size, head shapes, and skin colors", his qualifier "among human groups" is to be taken very seriously.[12] But then, what makes these "human groups" *human* if they cannot identify with some broader cultural/civilizational identity that would distinguish them from technically developed non-humans?

Huntington believes in strengthening the transatlantic relationship by cementing NATO into an Atlantic Community defined by its shared cultural heritage from Greece and Rome through the Renaissance, and its shared belief in the rule of law, parliamentary liberal democracy, capitalism and free trade.[13] Still, this universal state of Western Civilization would be that of only one civilization among others, though he sees it as a way to ensure that the West remains the strongest of them, especially as a consequence of establishing a common North Atlantic currency. He

---

11   Ibid., 43 my emphasis.

12   Ibid., 42.

13   Ibid., 307.

quotes Prime Minister Mahathir's warning to Asians that "With their trading clout, the EU-NAFTA confederation could dictate terms to the rest of the world."[14] In his view, whether or not the political and economic unification of Europe and North America occurs, depends above all on whether the United States reaffirms its Western identity and its leading role in Western Civilization.[15]

Huntington thinks that instead of deluding themselves into believing their values are universal, Westerners should take pride in the uniqueness of western culture, reaffirming, *preserving and protecting* our values from internal decay.[16] (This is a kind of conservatism that imagines 'Western values' to be static.) If others are to emulate the West by the adoption of European values such as individualism, social and political liberalism, and free enterprise that will be a credit to the West, but it must be voluntary. To impose these values on the world would require another phase of Western imperialist expansion, and global empire is, in Huntington's view, impractical.[17] To attempt to pursue it in the face of the declining power of a matured Western Civilization (i.e. effectively one in its old age) would mean war with the core states of other civilizations, such as China, and endanger the very survival of the West and its cherished ideals.[18] He thinks that instead an Atlanticist policy should be pursued to the end of unifying Western Civilization into a single universal state.[19]

While the values of no one civilization, including Western civilization, are universal, Huntington accepts that there are certain minimal commonalities and overlaps between human communities and it is through the exploration and expansion of these that over the very long run, humans might gradually develop a universal civilization.[20] *But we simply do not*

---

14 Ibid., 308.

15 Ibid.

16 Ibid., 310–11, 318.

17 Ibid., 310–311, 318.

18 Ibid., 311.

19 Ibid., 312.

20 Ibid., 318, 320.

have time for such a gradual development in the face of increasingly rapid and convergent advancements in technology that threaten the extinction of the human species and demand global decision making, based on a universal value system or worldview. In point of fact, the most universal worldview with any real ethical content is that of the kindred Indo-European cultures, which continued to deeply influence each other throughout the course of history, through the East-West nexus of *Irân-Shahr*.

The reverence for Wisdom, cultivation of the intellect and the contemplation of cosmic order as the criterion of humanization, ferocious truthfulness, aristocratic meritocracy and the unequivocal rejection of mob rule, chivalry and charitable free spiritedness, joyousness and an ecstatic self-overcoming of need and greed, industriousness and divinization of our own creative potential, and at the same time a recognition that respect for the Earth's ecology is a precondition for bodily health, vitality, and spiritual wholeness — these are some of the archaic Indo-European core values of a future, united Aryan world society. When I say future, I mean near future. It is these Indo-Iranian or Eastern Aryan values that are also co-constitutive of European civilization, which must form the bedrock of the constitutional order of the world state that emerges from out of the planetary emergency of the technological apocalypse. The Third World War may have begun as a clash of civilizations, but its proper end or *telos* is the constitution of an Indo-European World Order based on these fundamental principles.

Only such a development can provide us, not with a merely procedural world government, but with the kind of organically integrated world society that is capable of enduring the most profoundly challenging dimensions of convergent advancements in technology. For example, only in a society wherein the value of truthfulness is so deeply rooted will the inevitable total loss of privacy be something that we can cope with. The emphasis on industrious pursuit of progressive self-perfection will encourage the best uses of neo-eugenic biotechnologies, while the values of chivalric charitableness and respect for the free choice of the individual, will assure that genetic engineering is accessible to all and is not abused in a way that spawns a specialized slave race or a caste system reinforced by genetic design. The ecological dimension of Zarathustra's teaching would also obviously have a deep and broad impact on the way in which the Indo-European world state guides us through the technological singularity.

We have about 30 years to bring such a world state into being. That means it will be an incredibly violent, revolutionary process. Evolution is often contrasted with revolution by implicitly pacifistic moderates, but they forget that evolution entails not just mutation but the extinction of those who fail to make the leap into a new form of life. World War I resulted in 17 million deaths, and World War II in 70 million. Most of those killed were civilians. I think that we should consider ourselves very fortunate if the Third World War, the war for the Indo-European World Order in the face of the threat of an Islamic or Chinese world state emerging from out of the technological apocalypse, ends up claiming the lives of 700 million people. I think it is far more likely that it will end with around 700 million people left on this planet. In other words, I could easily see a reduction of Earth's population by 90% within a single generation.

Those left will not be human beings. They will be forerunners of a new species, which Nietzsche called the Superman and which he claimed would be as different from Man as Man is from the Ape. The *world state of emergency* is the concrete historical context for the fulfillment of this prophecy uttered by the returned Zarathustra, the first and greatest prophet of the Aryans. Those self-styled "identitarians" who want to hold on to Traditional Christianity and hole up in one of many segregated ethno-states, are, as we say in Persian, "not even in the outer courtyard" (*hatâ dar bâgh ham nistand*) let alone the inner sanctum of the new House of Being that Indo-Europeans will built from their ethno-linguistic consciousness.[21] Unless they see the light soon, they will perish together with the other *untermenschen*.

An archeo-futuristic development of a terminally declining Western Civilization, by means of integrating a Neo-Zoroastrian Greater Iran, might revitalize 'the West' by transforming it into a global Indo-European

---

21 Referencing Heidegger's statement, "Language in the house of Being."

or Aryan Civilization. Samuel Huntington is wrong that the very concept of a "universal civilization" is a unique product of Western Civilization.[22] The Persian Empire or *Irân-Shahr* also had the same conception of itself. The idea of a global civilization is not European, but Indo-European or Aryan in origin. It is because the mindset that unleashes unbounded technological-scientific transformations of the environment and every dimension of the human experience, which begins with Zarathustra, is also one that inevitably forces a common destiny on all cultures that are touched by its globalizing force.

Civilizations are the most enduring bases for identity. Civilizations outlast cultural, political, and ideological upheavals that take place within them — often involving warfare and diplomacy among sub-civilizational elements that is more intense than the relationship between any of these elements and cultures or countries from another civilization. In the *Clash of Civilizations*, Samuel Huntington claims that the measure of the transgenerational endurance of civilizations (as opposed to ideologically based political regimes or local cultures) is "certain primary structuring ideas."[23] But he also admits that this endurance should not be taken to mean that civilizations are static. Civilizations not only rise and fall in a set configuration, it is also the case that "they merge and divide."[24] What he fails to recognize is that there is something about the "primary structuring ideas" of Western Civilization that endows it with a unique world-conquering destiny, and that this is also something that connects the West to the wider Indo-European world — especially via Greater Iran. The unique destiny of the West has everything to do with what Huntington characterizes as the relationship between civilization in the singular — i.e. technical development — and civilizations in the plural.

Huntington takes civilization and culture to be basically synonymous. He acknowledges the opposing view of predominately German historians and social scientists that cultures are primitive, static, nonurban

---

22 Ibid., 66.

23 Huntington, *The Clash of Civilizations*, 43.

24 Ibid., 44.

communities defined by distinct values and customs, whereas civilization presupposes the evolutionary dynamism required for the development of technology and attendant urbanization. Against these scholars, with whom I agree, Huntington maintains that "a civilization is a cultural entity" or that civilization is simply "a culture writ large", in which what is definitive is not necessarily race or ethnicity but values, belief systems, social structures and institutions that sustain them across generations. The rejection of the view that Huntington associates with German scholars is extremely significant and consequential. Implicit in the rejected distinction between a technological civilization and a primitive barbarian culture is a potential basis for the ultimate convergence of civilizations.

China and Islam can threaten to dominate the Earth through the very same technological forces that we have empowered them with, but they will never understand what it is that they are even doing. We cannot trust the Chinese and the Muslims to make responsible decisions regarding, for example, how biotechnology can be used to enhance rather than degrade human life or at what point the integration of human consciousness with drone robotic systems becomes inhuman. This is our sovereign responsibility, and one that we will have to accept in the midst of a horrific war for planetary hegemony.

Consequently, *emergency* in my concept of a *world state of emergency* should also evoke *emergence* in the sense of Heraclitean and Nietzschean becoming or creative destruction through dynamic opposition, which was addressed in Chapter 2 in the context of Heidegger's quote of the 53rd fragment in his letter to Carl Schmitt. In other words, especially in view of its emergence from out of the sovereign decision made in a state of exception, when legal norms are suspended, the structure of this World State will have nothing essential in common with the liberal conception of global order. Hegemony does not entail an entirely pacified universal state. A hegemonic world state is still one sustained by strife and defined against what is outside — but what must also remain the losing side. Its constitution will be grounded on a definite ethos, and establish a planetary sphere of hegemonic power in defense of this ethos.

Attention to the specific challenges posed by convergent advancements in technology ought to drive home how in every case, this form of worldwide sovereign power is no mere whim or arbitrary prejudice, but rather an absolute necessity for safeguarding human existence as we face a unique evolutionary challenge. This demands an authoritarian and organic state, an Indo-European World Order centralized and hierarchically unified by a visionary and decisive sovereign. If we were to render his title in Persian, it would be *Âryâmehr Shâhanshâhé Irân va Anirân, Qebleyé Âlam va Lessân al-Qeyb*: "Light of the Aryans, King of Kings of the Aryan and Non-Aryan Realms, Axis of the Cosmos and Tongue of the Invisible."

Bibliography

- Afnan, Ruhi (1969) *Zoroaster's Influence on Anaxagoras, the Greek Tragedians and Socrates* (New York: Philosophical Library).

- Altheim, F. and R. Stiehl (1961) *Die aramäische Sprache unter den Achaimeniden* (Frankfurt).

- Asher, Catherine B. (1992) *Architecture of Mughal India* (New York: Cambridge University Press).

- Benoist, Alain de (2016) *The Indo-Europeans: In Search of the Homeland* (London: Arktos).

- Benoist, Alain de (2017) *View from the Right, Vol. 1* (London: Arktos).

- Blascovich, Jim, and Jeremy Bailenson (2011) *Infinite Reality: The Hidden Blueprint of Our Virtual Lives* (New York: William Morrow).

- Bodhi, Bhikkhu (2005) *In the Buddha's Words: An Anthology of Discourses from the Pali Canon* (Somerville, MA: Wisdom Publications, 2005).

- Brosius, Maria (1998) *Women in Ancient Persia* (New York: Oxford University Press).

- Corbin, Henry (1976) "From Heidegger to Suhrawardi", an interview with Philip Nemo (recorded for Radio France-Culture on June 2).

- Corbin, Henry (1977) *Spiritual Body and Celestial Earth* (Princeton: Princeton University Press).

- Corbin, Henry (1998) *The Voyage and the Messenger: Iran and Philosophy* (Berkeley: North Atlantic Books).

- Danner, Victor (1999) "Arabic Literature in Iran" in *The Cambridge History of Iran: Volume 4, From the Arab Invasion to the Saljuqs* (New York: Cambridge University Press).

- Davies, J.K. (1993) *Democracy and Classical Greece* (Cambridge: Harvard University Press).

- Davis, Dick (2008) "Introduction" in Fakhraddin Gorgani, *Vis and Ramin* (New York: Penguin Classics).

- Dawood, N.J. (translator) (1995) *The Koran* (New York: Penguin Classics).

- Dowman, Keith (1996) *Sky Dancer: The Secret Life and Songs of the Lady Yeshe Tsogyel* (Ithaca, NY: Snow Lion Publications).

- Dugin, Alexander (2017) *The Rise of the Fourth Political Theory* (London: Arktos).

- Edou, Jérôme (1996) *Machig Labdrön and the Foundations of Chöd* (Ithaca, NY: Snow Lion Publications).

- Emmerick, R.E. (1996) "Buddhism Among Iranian Peoples" in *The Cambridge History of Iran, Volume 3(2)* (New York: Cambridge University Press).

- Ernst, Carl W. (2003) "Muslim Studies of Hinduism? A Reconsideration of Arabic and Persian Translations from Indian Languages" in *Iranian Studies*, Vol. 36, No. 2, 173–195.

- Faye, Emmanuel (2009) *Heidegger: The Introduction of Nazism into Philosophy* (New Haven: Yale University Press).

- Frye, Richard (1993) *The Heritage of Persia* (Costa Mesa: Mazda Publishers).

- Fukuyama, Francis (2002) *Our Posthuman Future* (New York: Picador).

- Ghani, Muhammad Abdul (1972) *History of Persian Language and Literature at the Mughal Court: With a Brief Survey of the Growth of Urdu Language* (Gregg Publishing).

- Ghirshman, Roman *Persian Art: Parthian and Sassanian Dynasties, 249 B.C.–A.D. 651* (New York: Golden Press, 1962).

- Gobineau, J.A. (1971) *The World of the Persians* (Genève: Editions Minerva).

- Guthrie, Kenneth Sylvan (1988) *The Pythagorean Sourcebook and Library* (Grand Rapids: Phanes Press).

- Hallaq, Wael B. (2005) *The Origins and Evolution of Islamic Law* (New York: Cambridge University Press).

- Herodotus (1987) *The Histories* (Chicago: University of Chicago Press).

- Huntington, Samuel P. (2003) *The Clash of Civilizations and the Remaking of World Order* (New York: Simon and Schuster).

- Insler, Stanley (1990) "The Love of Truth in Ancient Iran" in *An Introduction to the Gathas of Zarathustra* (Pittsburgh).

- Jafarey, Ali (1989) *The Gathas, Our Guide: The Thought-Provoking Divine Songs of Zarathustra* (Cypress, CA: Ushta Publications).

- Jørgensen, John J. (2005) "Bodhidharma" in *Inventing Hui-neng, the Sixth Patriarch* (Leiden: Brill Academic Publishers).

- Kahn, Charles H. (1999) *The Art and Thought of Heraclitus* (Cambridge: Cambridge University Press).

- Keddie, Nikki R. (1981) *Roots of Revolution: An Interpretive History of Modern Iran* (New Haven: Yale University Press).

- Kent, Roland G. (1953) *Old Persian: Grammar, Texts, Lexicon* (American Oriental Society).

- Kunstler, James Howard (2005) *The Long Emergency: Surviving the End of Oil, Climate Change, and Other Converging Catastrophes of the Twenty-First Century* (New York: Atlantic Monthly Press).

- Littleton, C. Scott and Linda A. Malcor (2000) *From Scythia to Camelot* (New York: Routledge).

- Locke, John (1983) *A Letter Concerning Toleration* (Indianapolis: Hackett).

- Lynn, Richard (2001) *Eugenics: A Reassessment* (Westport: Praeger).

- Mack, Burton L. (1995) *Who Wrote the New Testament* (New York: Harper Collins).

- Mayer, Elizabeth Ann (2007) *Islam and Human Rights* (Boulder: Westview Press).

- Mill, John Stuart (1956) *On Liberty* (New York: Bobbs-Merrill).

- Montesquieu, Charles-Louis de Secondat (1984) *Persian Letters* (New York: Penguin Classics).

- Morsink, Johannes (1999) *The Universal Declaration of Human Rights: Origins, Drafting, and Intent* (Philadelphia: University of Pennsylvania Press).

- Naam, Ramez (2005) *More Than Human* (New York: Random House).

- Nabarz, Payam (2005) *The Mysteries of Mithras: The Pagan Belief That Shaped the Christian World* (Rochester, VT: Inner Traditions).

- Nanavutty, Piloo (1999) *The Gathas of Zarathustra: Hymns in Praise of Wisdom* (Ahmedabad: Mapin Publishing).

- Nasr, Seyyed Hossein (1999) "Bundahisn" and "Greater Bundahisn" in *An Anthology of Philosophy in Persia* (New York: Oxford University Press).

- Nietzsche, Friedrich (2000) *Basic Writings of Nietzsche* (New York: The Modern Library).

- Pahlavi, Reza (2002) *Winds of Change: The Future of Democracy in Iran* (New York: Regnery Publishing).

- Plato (1991) *The Republic of Plato* (Basic Books).

- Plato (1999) *Collected Dialogues* (Princeton: Princeton University Press).

- Quiles, Carlos and Fernando López-Menchero, *A Grammar of Modern Indo-European: Prometheus Edition* (2012).

- Rawls, John (1999) *A Theory of Justice* (New York: Oxford University Press).

- Rousseau, Jean-Jacques (1987) *The Basic Political Writings* (Indianapolis: Hackett).

- Sastri, Makhanlal Roychoudhury (2014) *The Din-i-Ilahi or the Religion of Akbar* (Nabu Press).

- Shamel, Shafiq (2013) *Goethe and Hafez: Poetry and History in the West-östlicher Divan* (New York: Peter Lang International Academic Publishers).

- Schmitt, Carl (1988) *The Crisis of Parliamentary Democracy* (New York: MIT Press).

- Schmitt, Carl (1996) *The Concept of the Political* (Chicago: University of Chicago Press).

- Schmitt, Carl (2006) *Political Theology* (Chicago: University of Chicago Press).

- Schmitt, Carl (2007) *Theory of the Partisan* (New York: Telos Press).

- Schmitt, Harrison (2010) *Return to the Moon: Exploration, Enterprise, and Energy in the Human Settlement of Space* (New York: Copernicus Books).

- Singer, P.W. (2010) *Wired for War: The Robotics Revolution and Conflict in the 21st Century* (New York: Penguin Books).

- Strauss, Leo (1965) *Natural Right and History* (Chicago: University of Chicago Press).

- Tafazzoli, A. and A.L. Khromov (1999) "Sasanian Iran: Intellectual Life" in *History of Civilizations of Central Asia: Volume III The Crossroads of Civilizations A.D. 250 to 750* (Paris: United Nations Educational, Scientific and Cultural Organization).

- Ulansey, David (1989) *The Origins of the Mithraic Mysteries* (New York: Oxford University Press).

- Voltaire, Francois (1984) "Zoroaster" in *Philosophical Dictionary* (New York: Penguin Classics).

- Watson, William (2003) "Iran and China" in *The Cambridge History of Iran: Volume 3(1), The Seleucid, Parthian and Sasanian Periods* (New York: Cambridge University Press).

- West, M.L. (2015) *The Hymns of Zarathustra* (New York: I.B. Taurus).

- Whitfield, Roderick and Susan Whitfield, and Neville Agnew *Cave Temples of Mogao: Art and History on the Silk Road* (Los Angeles: The J. Paul Getty Trust, 2000).

- Xenophon (2001) *The Education of Cyrus* (Cornell: Cornell University Press).

- Zarghamee, Reza (2013) *Discovering Cyrus: The Persian Conqueror Astride the Ancient World* (Washington: Mage Publishers).

- Zimmer, Heinrich (1968) *The Art of Indian Asia: Its Mythology and Transformations* (Princeton: Princeton University Press).

Index

A

Âbâdsâzié Giti 160
abortion 78, 92–94
Achaemenid 161–175, 192
agent 30, 120–121
Ahurâ 172, 191
Anarchism 61
apocalypse vii, xix–xx, 67, 155–159, 202–203
Arabia ix, 4, 18–22
Artificial Intelligence xv, 113
A.I. 113–122
Alans 176–183
Amordâd 159
Antarctic Treaty 144
Antarctic Convention 144
Antarctic Protocol 144
Arash 190–191
Aristocracy 166
Aristokratiâ 172
Ârtâ (see also *Ashâ*) 157–170
Artemis 169–170, 184
Aryamehr 26
Aryan vii–xx, 74, 153–169, 183–206
Ashâ (see also *Ârtâ*) 168–170
Asian 2–12, 74, 91–98, 155–161, 189
authoritarian 61–73, 90–96, 206

avatar 120–121, 179
Avestâ 178
Avicenna 197

B

Baghdad 196
Bahman 157–166
Biotechnology vii–xii
Biruni 197
BrainGate 114
Bodhidharma 188, 209
Bozorgmehr 161, 196
Buddha viii, 166–195, 207
Buddhism viii, 154–192, 208

C

Caliphate viii–xx, 4, 55, 196–199
Cambrian explosion 102
Caucasus xix, 13, 154, 176–181, 197
Central Asia xx, 6–10, 96, 181, 196, 211
Chernobyl 137
China xvii–xx, 2–14, 32, 72, 91–103, 139–147, 164, 189, 201–212
Chinese viii–xii, 2–14, 69–71, 92–96, 155, 189–205
Chivalry 176–177
Christianity x, 3, 16–24, 54–55, 177–186, 203
Clash of Civilizations 1, 199–209

cloning xiii, 69–78, 93–104, 153
Confucian 2–13
Corbin 193–194, 207
Cosmopolitan 5–10, 164
Cyberspace 113–122, 153
Cyrus 161–190, 212

D
Daevâ 177–178, 191
Daheshmandi 166
Darius 158–170
DARPA xiv, 110–115
Darwin 71–72
Democracy vii–xx, 1–41, 53–56, 146–153, 166–172, 200–211
DNA 80–87, 101–103
Doogie Mice 85
drone xiv–xv, 107–124, 205
dysgenic 72–93

E
embryonic selection 69–80, 95–99
Enlightenment ix, 19, 32, 155–159, 188–194
EU 93–97, 201
Eugenics xii, 69–99, 210
Europe vii, xx, 2–13, 55, 73–75, 91–92, 129–137, 155–156, 176–201
European vii–xx, 5–14, 73–77, 93–97, 141, 154–211
European International System 5
existential vii–xi, 50–64

F
Farabi 197
Farvahar 198
Farré kiâni 161
Fascist 3–13

Ferdowsi 157, 183–186
Frashgard 159
fravashi 157, 172
free radicals 87
Fukushima 137
Fukuyama 77–101, 208
Fusion 140

G
Galton 70–74
gasoline 128–135
general will ix, 50–56
genetic engineering xii–xiii, 69–101, 153, 202
Goethe 193, 211
Gondeshapur 195–196
Gorgan 181
Gorgani 185–193, 208
Goshtasp 161–172, 195
Goth 182
Greece ix, 36–41, 53–59, 156–171, 196–208
Greek vii, 36–38, 53–76, 154–207

H
Hegel 5, 159, 193
hegemonic 8, 205
hegemony viii–xix, 7, 205
Heidegger 56–60, 194–208
Helium-3 xvi–xviii, 124–125, 138–150
Helsinki Declaration 90
Heraclitus 57–60, 158–170, 209
Herodotus 38, 161–167, 209
Homeric 173
Hubbert 125–127
human rights vii–xx, 2–33, 89, 153–163
human nature 29, 51

INDEX

hybrid xiii, 35, 101
hybridization 104
hydroelectric 132–135
hydrogen fuel cell 134–135

I

IAEA 89
India vii, 12, 97, 139–171, 183–207
Indo-European vii-xx, 154–160, 177–211
Iran vii, xix-xx, 3–12, 26–28, 153–212
Iranian xix-xx, 26–27, 154–208
Irân-Shahr xix-xx, 192–204
Islam ix-xx, 3–35, 54–55, 155, 183–210
Islamic viii-xx, 2–28, 55, 126, 155, 185–209
IQ xii, 80–104

J

Japan 5–12, 95–97, 129–147, 189
Javânmardi 176–189
Jews 16, 74, 165, 184
Judaism 16–24

K

Kanishka 187–188
Khayyam 194–197
Khorasan xx, 186–197
Khordâd 158
Khosrow 161, 195
Khwarazmi 197
Kurdistan 156, 177, 198
Kurds xx, 156, 198–199
Kushans 158, 186–192

L

Law of the Sea 145–149
Liberal x, 35–40
Liberalism ix, 35–54

Liberty 15, 35–50, 210
LNG 131
Locke 15–16, 210
Lynn 69–100, 210

M

Magi 167–180
Mahayana 161, 177–191
Maitreyâ 187
Mani 161, 189–195
Marxism 3–5
Mazdâ 157, 170–172, 191
Middle East 13, 137
Mill x, 35–54, 210
Mithra 158, 177–198
Mithraism 158–161, 177–189
Moon vii-xviii, 124–125, 138–151, 211
Moon Agreement 146–149
Mughal 191–192, 207–208
Muhammad 17–25, 54–55, 191–208
Multiculturalism 11
Muslim viii-ix, 3, 22–24, 138, 192–196, 208

N

natural gas 129–139
Nietzsche 56–59, 170, 193–210
noble lie 173–174
NAFTA 130, 201
North America 11–14, 92, 130, 201
North Atlantic Treaty Organization (NATO) xi, 92, 200
NPT 89
nuclear xi-xvii, 13, 66–69, 87–90, 129–143
Nuremberg Code 90

O

Oligarchy 172
OPEC 2, 126–129
Ordibehesht 157–166
Ossetia 156
Outer Space Treaty 147–149

P

Pahlavi 26, 186–196, 210
Pahlavân 186
Pakistan 4–13, 90, 111, 191–192
Paridaezâ 160
partisan xi, 65–66
peak oil vii–xvi, 125–132
Perikles 35
Persia 55, 155–187, 207–210
Persians xx, 38, 55, 156–209
Persianate vii, 166, 191–197
Persian Gulf xvii, 124–126, 151–153
planetary emergency vii, xix, 67, 202
Plato 5, 39–45, 69–70, 157–176, 195–196, 210
Pre-implantation genetic diagnosis (PGD) 78, 97
Progress 5–19, 32, 65–70, 95–105
Pythagoras 166–176

Q

qanât 160
Quran 16–32

R

Rawls 92–93, 211
Razi 197
Renaissance xx, 195–200
robot xiv, 107–113
Romans 8, 156, 179–190
Rome 54, 155, 181–185, 200
Rostam 183
Rousseau ix–x, 35, 51–56, 211
Russia xvii, 4–14, 74, 96, 137–147

S

Sanskrit 154, 177–196
Sarmatian 156, 176–184
Saudi 2–4, 18–23
Schmitt x–xix, 35–41, 55–66, 125, 138–155, 205–211
Scythian 158, 176–197
secular ix–x, 3–10, 54–56
Sepandârmad 158–159
Sepandminou 159–160
Shahanshah 26, 181
Shâhnâmeh 177–186
Shahrivar 160, 188
Shanghai Cooperation Organization viii, 6–8, 96
Shapur 161, 195
sharia 17, 55
singularity xx, 67, 203
Socrates 39–56, 167–173, 207
solar panels 132
South Korea 5, 77, 91–95, 120, 139
sovereign x–xix, 6–15, 28, 42–67, 110, 146–151, 205–206
state of emergency vii–xx, 62–67, 203–205
state of exception x, 62, 205
Strauss 54–61, 193, 211
Suhrawardi 191–194, 207

T

Tajik 156
Tâj Mahal 192
Third World War viii–xix, 1–2, 27–28, 202–203

Three Mile Island 137
Tolerance 155
Toleration 15, 210
Transformers 110
Transhumanist 104
Tyr 190–191

U

Ukraine xix, 154–156, 190
United Nations viii–ix, 14, 26–28, 76, 89–97, 145, 196, 211
United States 2–12, 28–41, 60, 73–74, 90–99, 111–112, 126–150, 163, 199–201
Universal Declaration of Human Rights (UDHR) viii, 2, 14–34, 64, 210

V

Vajrâyânâ 191
Virâ 191
Virtual Reality 107–123
Vohuman (see also *Bahman*) 158–167

W

Western Civilization 8–14, 155–156, 200–204
women ix, 18–37, 71–94, 159–198
world society xviii, 122, 153, 202
world state vii–xx, 64–69, 98–104, 122, 153–155, 202–205
world war 1–3, 27–28, 155

X

Xerxes 161–167
Xvarnah (see *farré kiâni*) 161

Y

yazâtâ 178
Yazidi 198

Z

Zarathustra 157–212
Zoroaster 167–177, 194, 207–212
Zoroastrianism 155–172, 189–198

OTHER BOOKS PUBLISHED BY ARKTOS

Sri Dharma Pravartaka Acharya	*The Dharma Manifesto*
Alain de Benoist	*Beyond Human Rights*
	Carl Schmitt Today
	The Indo-Europeans
	Manifesto for a European Renaissance
	On the Brink of the Abyss
	The Problem of Democracy
	View from the Right (vol. 1–3)
Arthur Moeller van den Bruck	*Germany's Third Empire*
Matt Battaglioli	*The Consequences of Equality*
Kerry Bolton	*Revolution from Above*
Isac Boman	*Money Power*
Ricardo Duchesne	*Faustian Man in a Multicultural Age*
Alexander Dugin	*Eurasian Mission*
	The Fourth Political Theory
	Last War of the World-Island
	Putin vs Putin
	The Rise of the Fourth Political Theory
Koenraad Elst	*Return of the Swastika*
Julius Evola	*Fascism Viewed from the Right*
	A Handbook for Right-Wing Youth
	Metaphysics of War
	Notes on the Third Reich
	The Path of Cinnabar
	A Traditionalist Confronts Fascism
Guillaume Faye	*Archeofuturism*
	Archeofuturism 2.0
	The Colonisation of Europe
	Convergence of Catastrophes
	Sex and Deviance
	Understanding Islam
	Why We Fight
Daniel S. Forrest	*Suprahumanism*
Andrew Fraser	*Dissident Dispatches*
	The WASP Question

OTHER BOOKS PUBLISHED BY ARKTOS

Génération Identitaire	*We are Generation Identity*
Paul Gottfried	*War and Democracy*
Porus Homi Havewala	*The Saga of the Aryan Race*
Rachel Haywire	*The New Reaction*
Lars Holger Holm	*Hiding in Broad Daylight*
	Homo Maximus
	Incidents of Travel in Latin America
	The Owls of Afrasiab
Alexander Jacob	*De Naturae Natura*
Jason Reza Jorjani	*Prometheus and Atlas*
Roderick Kaine	*Smart and SeXy*
Lance Kennedy	*Supranational Union and New Medievalism*
Peter King	*Here and Now*
	Keeping Things Close
Ludwig Klages	*The Biocentric Worldview*
	Cosmogonic Reflections
Pierre Krebs	*Fighting for the Essence*
Stephen Pax Leonard	*Travels in Cultural Nihilism*
Pentti Linkola	*Can Life Prevail?*
H. P. Lovecraft	*The Conservative*
Charles Maurras	*The Future of the Intelligentsia & For a French Awakening*
Michael O'Meara	*Guillaume Faye and the Battle of Europe*
	New Culture, New Right
Brian Anse Patrick	*The NRA and the Media*
	Rise of the Anti-Media
	The Ten Commandments of Propaganda
	Zombology
Tito Perdue	*Morning Crafts*
	William's House (vol. 1–4)
Raido	*A Handbook of Traditional Living*

OTHER BOOKS PUBLISHED BY ARKTOS

STEVEN J. ROSEN	*The Agni and the Ecstasy*
	The Jedi in the Lotus
RICHARD RUDGLEY	*Barbarians*
	Essential Substances
	Wildest Dreams
ERNST VON SALOMON	*It Cannot Be Stormed*
	The Outlaws
SRI SRI RAVI SHANKAR	*Celebrating Silence*
	Know Your Child
	Management Mantras
	Patanjali Yoga Sutras
	Secrets of Relationships
TROY SOUTHGATE	*Tradition & Revolution*
OSWALD SPENGLER	*Man and Technics*
TOMISLAV SUNIC	*Against Democracy and Equality*
ABIR TAHA	*Defining Terrorism: The End of Double Standards*
	The Epic of Arya (2nd ed.)
	Nietzsche's Coming God, or the Redemption of the Divine
	Verses of Light
BAL GANGADHAR TILAK	*The Arctic Home in the Vedas*
DOMINIQUE VENNER	*For a Positive Critique*
	The Shock of History
MARKUS WILLINGER	*A Europe of Nations*
	Generation Identity
DAVID J. WINGFIELD (ED.)	*The Initiate: Journal of Traditional Studies*

Printed in Poland
by Amazon Fulfillment
Poland Sp. z o.o., Wrocław

24809999R00138